The Application of Ant Colony Optimization

Edited by Ali Soofastaei

Published in London, United Kingdom

IntechOpen

Supporting open minds since 2005

The Application of Ant Colony Optimization
http://dx.doi.org/10.5772/intechopen.91586
Edited by Ali Soofastaei

Contributors
Heoncheol Lee, Majid Vafaeipour, Dai-Duong Tran, Thomas Geury, Mohamed El Baghdadi, Omar Hegazy,
Martina Umlauft, Wilfried Elmenreich, Abhishek Kaul, Ali Soofastaei

Notice
Statements and opinions expressed in the chapters are these of the individual contributors and not
necessarily those of the editors or publisher. No responsibility is accepted for the accuracy of
information contained in the published chapters. The publisher assumes no responsibility for any
damage or injury to persons or property arising out of the use of any materials, instructions, methods
or ideas contained in the book.

First published in London, United Kingdom, 2022 by IntechOpen
IntechOpen is the global imprint of INTECHOPEN LIMITED, registered in England and Wales,
registration number: 11086078, 5 Princes Gate Court, London, SW7 2QJ, United Kingdom
Printed in Croatia

British Library Cataloguing-in-Publication Data
A catalogue record for this book is available from the British Library

Additional hard and PDF copies can be obtained from orders@intechopen.com

The Application of Ant Colony Optimization
Edited by Ali Soofastaei
p. cm.
Print ISBN 978-1-83968-176-9
Online ISBN 978-1-83968-177-6
eBook (PDF) ISBN 978-1-83968-178-3

Meet the editor

Dr. Ali Soofastaei is a global artificial intelligence (AI) projects leader, an international keynote speaker, and a professional author. He completed his Ph.D. and postdoctoral research fellowship at The University of Queensland, Australia, in the field of AI applications in mining engineering, where he led a revolution in the use of deep learning and AI methods to increase energy efficiency, reduce operation and maintenance costs, and reduce greenhouse gas emissions in surface mines. As a scientific supervisor, he has provided practical guidance to undergraduate and postgraduate students in mechanical and mining engineering and information technology for many years. Dr. Soofastaei has more than fifteen years of academic experience as an assistant professor and leader of global research activities. Results from his research and development projects have been published in international journals and keynote presentations. He has presented his practical achievements at conferences in the United States, Europe, Asia, and Australia. He has been involved in industrial research and development projects in several industries, including oil and gas (Royal Dutch Shell), steel (Danieli), and mining (BHP, Rio Tinto, Anglo American, and Vale). His extensive practical experience in the industry has equipped him to work with complex industrial problems in highly technical and multi-disciplinary teams.

Contents

Preface

Advanced analytics in both science and technology is growing rapidly, and optimization is essential for this growth. Applying innovative optimization approaches such as population-based metaheuristic models instead of using traditional models can help researchers obtain more benefits. In many fields and different domains of human activity, optimization problems are encountered frequently. As a result, we must find optimal or near-optimal solutions for specific issues to meet certain constraints. More specifically, optimization is concerned with the development of efficient and reliable computing infrastructures, which will be used, among other things, to accelerate meta-heuristic techniques by significantly improving their performance. Numerous heuristic algorithms have been developed to find faster, near-optimal solutions to reduce time to market.

Moreover, heuristic algorithms can quickly generate a solution with acceptable quality. Ant Colony Optimization (ACO) is one the most critical and widely used models among heuristics and meta-heuristics, including genetic algorithms, Simulated Annealing, and Gray Wolf Optimization. This book provides an overview of ACO applications in various fields as well as their technical details.

Structure of the book

This book discusses four ACO technic applications in different industries. Practical examples and scientific details support all the information. The chapters contain enough information for beginners to familiarize themselves with the high technology and science application to solve business problems and more detailed technical information for advanced readers.

Chapter 1 provides a background of ACO analytical model application to help industries make better decisions to optimize processes and reduce costs.

Chapter 2 discusses the application of ACO for the integrated design of Hybrid Electric Vehicles (HEVs). It examines the actual application of continuous ACO for integrated sizing and control design of HEVs to minimize drivetrain cost fuel consumption and address control objectives. The chapter provides valuable information for designers and automotive engineers related to incorporating soft computing, modeling, and simulation concepts into the optimization-based design of HEVs.

The authors of Chapter 3 are members of a research group at the Department of Information Technology Convergence Engineering, School of Electronic Engineering, Kumoh National Institute of Technology, Korea. The researchers for multi-robot systems investigated merging grid maps with ACO. Multi-robot systems have recently come into the spotlight due to their efficiency in performing tasks in a collaborative environment. However, if there is no map in the working environment, each robot must complete SLAM, which is a process that simultaneously performs localization and mapping of the surrounding environment. To operate the multi-robot systems efficiently, the individual maps must be merged into a collective map that is accurate and complete. When the initial correspondences between the robots are unknown or uncertain, the map merging task becomes more challenging to complete. This chapter describes a novel approach to successfully and efficiently conducting grid-map merging with ACO, one of the well-known sampling-based optimization algorithms. This method was tested with one of the existing grid maps combining algorithms. The results showed that the ACO increased the accuracy of grid-map merging by approximately 20 percent.

In Chapter 4, the authors focus on ACO application for routing in wireless multi-hop networks. Wireless Mesh Networks (WMNs) and Mobile Ad-Hoc Networks (MANETs) are applied in situations where there is no predefined network structure consisting of routers and a base station or where the network is dynamic due to a growing number of nodes or mobile nodes moving into areas that a base station has not previously covered. This chapter introduces Wireless Multi-Hop Networks, their specific challenges, and an overview of the ACO application for routing in such networks.

Chapter 5 is written by an IT manager from IBM Singapore who has worked in advanced analytics applications in different industries for many years. The chapter focuses on preventive, predictive maintenance using ACO. The presented study results of using ACO to reduce maintenance costs in the mining industry can be a compelling case for the researchers who are thinking about a successful example of using advanced analytics to reduce maintenance costs.

We hope this book helps readers, including industry professionals and researchers, better understand ACO model applications in different areas. The chapters in this book present the state of the art of critical topics in ACO. Furthermore, each section's breadth of coverage and depth make it a helpful resource for all managers and engineers interested in the new generation of data analytics applications. Above all, the editor hopes that this volume will spur further discussions on all aspects of ACO application in different industries.

Dr. Ali Soofastaei
Artificial Intelligence Center,
Vale, Brisbane, Australia

Introductory Chapter: Ant Colony Optimization

Ali Soofastaei

1. Introduction

Optimization challenges arise in a wide range of fields and sectors of human activity, where we must discover optimal or near-optimal solutions to specific problems while staying within certain constraints. Optimization issues are essential in both scientific and industrial fields. Timetable scheduling, traveling salesman problems, nurse time supply planning, railway programming, space planning, vehicle routing problems, Group-shop organizing problems, portfolio improvement, and so on are a number of real-world illustrations of optimization opportunities. For this reason, many optimization algorithms are created [1].

Optimization focuses on establishing efficient and reliable computational infrastructures that will be used, among other things, to improve the performance of meta-heuristic techniques dramatically. As a result, numerous heuristic algorithms for identifying speedier near-optimal solutions have been created. Moreover, heuristic algorithms can solve with acceptable quality in a short amount of time [2].

Scientists have devoted a great deal of work to understand the complex social habits of ants, and computer scientists are now learning that these patterns can be exploited to tackle complex combinatorial improvement challenges. Ant colony optimization (ACO), the most successful and generally recognized algorithmic technique based on ant behavior, results from an effort to design algorithms motivated by one element of ant behavior, the capability to locate what computer scientists would term shortest pathways. ACO is a population-based metaheuristic for resolving complex optimization challenges. This method is a probabilistic optimization procedure used to solve computational issues and discover the best path using graphs. Artificial ants in ACO seek software agents for possible answers to a particular improvement issue. The optimization challenge is the challenge of discovering the optimum path on a weighted diagram to use the ACO. The artificial ants (hence referred to as ants) then incrementally create solutions by traveling along the graph. Using the pheromone model, a set of parameters associated with graph components (nodes or edges) whose values are updated at runtime by the ants, the solution construction process is skewed in one direction [3].

ACO is a well-known bio-inspired combinatorial optimization approach. Marco Dorigo proposed ACO in his Ph.D. thesis in the early 1990s to solve the optimal path issue in a graph [4]. It was first used to resolve the well-known dilemma of the traveling salesman, and it has since become widely used. After that, it's used to solve various complex optimization problems of several types. A great deal of time has been spent studying the complex social habits of ants, and computer scientists are now discovering that similar patterns can be exploited to solve complex combinatorial optimization problems, which represents a significant advance in the field. Each cycle begins with a departure from the nest, searching for a food source, and ends with a return to the nest. Each ant leaves a chemical known as pheromone on the path they walk during the journey. The pheromone concentration on each path is determined by the path's length and the quality of the accessible food supply. Because the concentration of pheromones present on a path affects ant selection, the higher the pheromone concentration, the more likely it is that ants will select the path. Using pheromone concentration and some heuristic value, such as the objective function value, each ant chooses a path in a probabilistic manner based on their environment [5].

Consider the following illustration. Let us consider the following scenario: there are two options for obtaining food from the colony. At first, there is no pheromone to be found on the ground. Consequently, the probability of choosing either of these two paths is equal, or 50 percent. For example, consider two ants who decide two alternative routes to obtain food, each with a fifty-fifty chance of success (see **Figure 1(a)**).

A significant amount of distance separates these two routes. Therefore, the ant who takes the shortest path to the food will be the first to reach it (see **Figure 1(b)**).

It returns to the colony after locating food and carrying some food with it. It leaves pheromone on the ground as it follows the returning path. The ant that takes the shortest route will arrive at the colony first (see **Figure 1(c)**).

As soon as the third ant decides to go out in search of food, it will choose the path that will take it the shortest distance, determined by the level of pheromones on the ground. A shorter road contains more pheromones than a longer path (see **Figure 1(d)**). The third ant will choose the shorter path because it is more convenient.

Upon returning to the colony, it was discovered that more ants had already traveled the path with higher pheromone levels than the ant who had taken the longer route. Therefore, when another ant tries to reach the colony's goal (food), it will discover that each trail has the same level of pheromones as the previous one. As a result, it selects one at random from the list. **Figure 1(e)** depicts an example of the option described above.

After several repetitions of this process, the shorter path has a higher pheromone level than the others and is more likely to be followed by the animal. As a result, all ants will take the shorter route the next time (see **Figure 1(f)**).

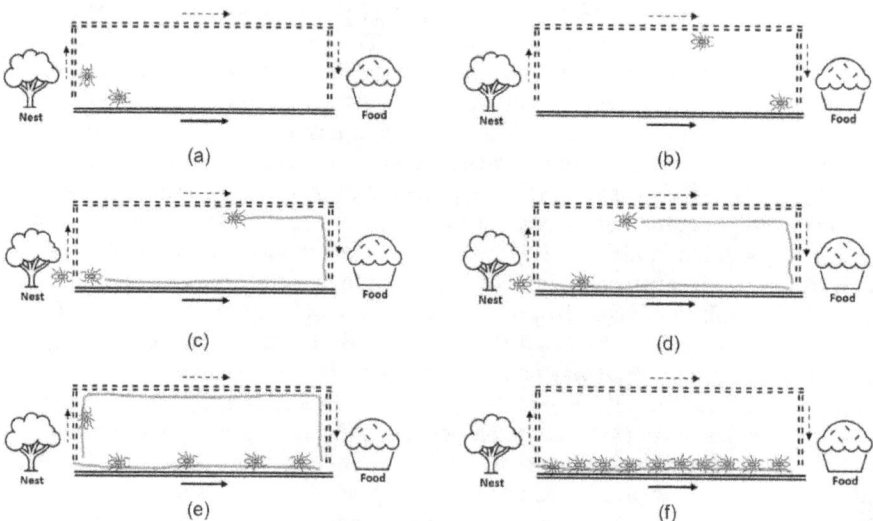

Figure 1.
Ant Colony optimization – A simple schematic view (a to f) [6].

An ACO is based on the technique known as Swarm Intelligence, which is a component of Artificial Intelligence (AI) methodologies for solving technical problems in the industrial sector [7].

2. Swarm intelligence

Amorphous computing comprises a large number of interconnected computers with low processing power, memory, and intercommunication modules. Amorphous computing is also referred to as distributed computing. Swarms are the collective name for these collections of electronic devices. When the individual agents interact locally, the computer's desired coherent global behavior results from the computer's local interactions. Although there are only a small number of misbehaving agents, and the environment is noisy and threatening, the global behavior of these enormous numbers of faulty agents is long-lasting. Therefore, randomness, repulsion, and unpredictability among agents can be used to derive Swarm Intelligence (SI), which can then be used to generate multiple solutions for a single problem. On the other hand, there are no established criteria for evaluating the performance of SIs [8].

On the other hand, SI is based on simple principles that allow it to solve complex problems with only a few simple agents. An SI feature causes coherent functional global patterns to emerge from the collective behaviors of (unsophisticated) agents interacting with their environment on a local level. SI provides a foundation for investigating collaborative (or dispersed) problem-solving approaches that do not rely on centralized control or an overarching model. SI refers to the natural or artificial behavior of decentralized, self-organized collective systems that operate on their initiative. The concept is commonly used in AI research and development. Since the early 1990s, a significant amount of effort has been expended on the solution of 'toy' and real-world problems using algorithms inspired by social insects [9].

Despite the fact that many studies on SI have been presented, there are no standard criteria for evaluating the performance of a SI system. As indexes, fault tolerance and local superiority are proposed. They used simulation to compare two SI systems in terms of these two indexes. There is a pressing need for additional analytical research.

According to the researchers, "continuum models" for swarm behavior should be based on nonlocal interactions found in biology. First, they discovered that when the density dependency of the repulsion term is greater than the density dependency of the attraction term, the swarm has a constant inner density with sharp edges, similar to what is observed in biological examples. Following that, they looked for linear stability at the swarm's borders [10].

2.1 Swarm intelligence: the fundamental principles

The following are the fundamental principles of swarm intelligence [11]:

1. Self-Organization is based on

   - positive feedback;

   - negative feedback;

   - amplification of fluctuations; and

   - multiple interactions.

2. Stigmergy- Indirect interaction through communication with the environment.

The purpose of this engagement is to provide a comprehensive review of the current state of the art in Swarm Intelligence, with a particular emphasis on the role of stigmergy in distributed problem-solving. The scope of this engagement is broad and includes a variety of topics. However, to proceed, it is necessary to provide working definitions and the essential properties of swarm-capable systems, such as the fact that problem-solving is an emergent property of a system of primary agents. The stigmergy concept states that simple agents can interact with one another over a common channel without a centralized control system. As a result of applying this concept, they are querying individual agents reveals little or nothing about the system's emergent characteristics [12].

Consequently, simulation is frequently used to understand better the emergent dynamics of stigmergic systems and their interactions. Individual acts in stigmergic systems are frequently selected from a restricted behavioral repertoire in a probabilistic manner. It is the activities of the various agents that cause changes in the environment, for example, the deposit of a volatile chemical known as a pheromone. Other agents are alerted to the presence of this chemical signal, which results in a shift in the probabilistic selection of future actions.

The advantages of a system like this are self-evident. Generally speaking, the activity of a single agent is less important in a system where the actions of several agents are required for a solution to emerge. Stigmaria systems are resilient to the failure of individual agents while also responding exceptionally well to dynamically changing contexts, as demonstrated in the following example. When developing algorithms, they are making the most efficient use of available resources is usually a significant consideration. One other type of stigmaria system, the raid army ant model, uses pheromone-based signaling to forage for food and survive efficiently. Agents in an army ant system establish a forage front covering a large area, resulting in extraordinarily successful food discovery. This model has military value because it could be used to develop a system for searching for land mines, which is a problem that is all too common in some parts of the world and that this model could help solve. This model of military interest is the third stigmaria model of military interest, characterized by flocking or aggregation. Many simple agents can be programmed to travel across an environment filled with obstacles (and potentially dangerous threats) without the need for centralized control or supervision. The agents' positions and velocities serve as cues to the environment they are operating.

2.2 Swarm intelligence advantages and disadvantages

There are several advantages of swarm Intelligence. In the following, some of them have been mentioned (**Table 1**) [13].

- Agents are simple, having little memory and behavior.

- Agents are not goal-oriented; rather than planning exhaustively, they respond.

- There is no central repository of information in the system, and control is dispersed.

- Agents are capable of reacting to rapidly changing settings.

Adaptable	The colony reacts to both internal and exterior disturbances.
Strong	Even though some individuals fail, tasks get completed.
Scalable	From a handful of individuals to tens of thousands of people
Distributed	In the colony, there is no such thing as a central controller.
Self-Organized	Paths to solutions are emergent rather than predefined

Table 1.
Advantages of swarm intelligence [13].

Behavior	Individual rules are challenging to predict collective behavior.
Knowledge	Interrogate one of the participants; it will not reveal anything about the group's function.
Sensitivity	Minor modifications in the rules result in a shift in group behavior.
Action	Individual action appears to be random: how can you spot threats?

Table 2.
Disadvantages of swarm intelligence [14].

- Individual agent failure is permitted, and emergent behavior is resilient to individual failure.

- It is not necessary to communicate with the agents directly.

Swarm intelligence does have some drawbacks. They are as follows (**Table 2**) [14].

- Individual agent behavior cannot be used to infer collective behavior. This means that watching solitary agents will not always lead to the selection of swarm-defeating behavior. (From an aggressive standpoint, this can be seen as a benefit.)

- Because action selection is stochastic, individual behavior appears to be noise.

- Swarm-based systems are challenging to design. For design, there are essentially no analytical mechanisms.

- Various factors influence the formation (or non-formation) of collective behavior differently.

3. Why is ants' behavior used for optimization?

What makes ants so fascinating?

- Ants use simple local methods to complete complex jobs.

- Ant's production exceeds the sum of their actions.

- Ants are experts at finding and exploiting resources.

Which ant mechanism is superior?

- Cooperation and division of labor are two terms that come to mind when discussing cooperation and labor.

- Adaptive job assignment.

- Cultivation stimulates work.

- Pheromones.

4. Applications of ACO

An increasing number of complex combinatorial optimization problems, such as quadratic assignment, fold protein folding, and vehicle routing, have been successfully solved with the aid of ACO algorithms in the past. Dynamic problems with real-world variables, stochastic problems, multiple targets, and parallel implementations have all been addressed by derived methods in various settings. It has also been employed to search for near-optimal solutions to the traveling salesman's problem. It is advantageous to use the ant colony algorithm when the graph changes dynamically because it can be performed continuously and adapt to real-time changes instead of simulated annealing and genetic algorithm techniques. This is relevant in network routing and urban transportation systems, among other areas.

Using ant colony optimization techniques, for example, it has been possible to find nearly optimal solutions to the traveling salesman problem. The Ant system, the world's first ACO algorithm, was created to solve the traveling salesman problem, which entails finding out which route is the most efficient between a set of locations. Essentially, the method is built around ants, who each embark on one of the numerous roundtrips while also visiting the various towns and cities. The ant decides how to travel from one city to another based on a set of guidelines that are followed at each stage [15]:

- Each city must be visited exactly once.

- A faraway city has a lower likelihood of being chosen (the visibility).

- The more intensive the pheromone trail laid out on the boundary between two cities, the more likely that edge will be chosen.

- If the travel is brief, the ant leaves more pheromones on all of the edges it passes through.

- Pheromone trails dissipate with each iteration.

In terms of applications, one of the hottest topics right now is the use of ACO to solve dynamic, stochastic, multi-objective, uninterrupted, and mixed-variable optimization problems and the development of parallel implementations that can make use of the latest similar technology.

ACO can be used to identify the best solution to various optimization challenges. Here are a few examples:

- Capacitated vehicle routing problem

- Permutation flow shop problem

- The issue of stochastic vehicle routing

- There is a problem with the vehicle routing for both pick-up and delivery

- Group-shop scheduling problem

- Traveling salesman problem

- The scheduling of nursing time is a complex problem

- Frequency assignment challenge

- Redundancy allocation problem

The application of ACO in different industries is growing very fast. In the near future, many newly developed ACP-based applications will be used in industries to improve the operational process.

5. Future of ACO

From this overview, it should be evident that the ACO metaheuristic is a robust foundation for tackling complicated combinatorial issues. Traveling Salesman Problem (TSP), Dynamic Traveling Salesman Problem (DTSP), Quadratic Assignment Problem (QAP), and other optimization challenges are particularly well suited to Ant Systems (AS). Compared to different powerful approaches, their superiority is undeniable, and the time savings (produced tours) paired with a high level of optimality cannot be overlooked in this context.

Ant System has a competitive advantage over highly communicative multi-agent systems thanks to reinforcement learning and greedy searching concepts. Simultaneously, the capacity to self-train fast enables light and agile installations. Finally, by separating individual swarm agents from one another, AS can process massive data volumes with far less waste than competing algorithms.

ACO is a new area that combines straightforward functions with profound conceptualizations. There's not much doubt that further research will yield exciting results that could provide answers to currently unsolvable combinatorial problems in the future. Alternatively, AS and ACO have been shown to be effective in various TSP permutations.

New research can provide better solutions by increasing effectiveness while decreasing restrictions by studying the ACO and PSO to make future improvements. More options for dynamically determining the optimal destination can be developed through ACO. For example, a plan to equip PSO with fitness sharing technology is being tested to see if it can help improve performance. In the future, rather than relying solely on the current iteration, each individual's velocity will be updated by combining the best elements from all previous iterations.

Author details

Ali Soofastaei
Artificial Intelligence Center, Vale, Brisbane, Australia

*Address all correspondence to: ali@soofastaei.net

IntechOpen

References

[1] Alkaya AF. Optimizing the Operations of Electronic Component Placement Machines. Turkey: Marmara Universitesi; 2009

[2] Blum C. Ant colony optimization: Introduction and recent trends. Physics of Life Reviews. 2005;2(4):353-373

[3] Chiong R, Neri F, McKay RI. Nature that breeds solutions. International Journal of Signs and Semiotic Systems (IJSSS). 2012;2(2):23-44

[4] Dorigo M, Di Caro G. Ant colony optimization: A new meta-heuristic. In: Proceedings of the 1999 Congress on Evolutionary Computation-CEC99 (Cat. No. 99TH8406). Washington DC, USA: IEEE; 1999

[5] Dorigo M, Birattari M, Stutzle T. Ant colony optimization. IEEE Computational Intelligence Magazine. 2006;1(4):28-39

[6] Dorigo M, Blum C. Ant colony optimization theory: A survey. Theoretical Computer Science. 2005;344(2-3):243-278

[7] Dorigo M, Socha K. An introduction to ant colony optimization. In: Handbook of Approximation Algorithms and Metaheuristics. Second ed. New York, USA: Chapman and Hall/CRC; 2018. pp. 395-408

[8] Dorigo M, Stützle T. Ant colony optimization: Overview and recent advances. In: Handbook of Metaheuristics. London: Springer; 2019. pp. 311-351

[9] Fattahi P et al. Sequencing mixed-model assembly lines by considering feeding lines. The International Journal of Advanced Manufacturing Technology. 2012;61(5):677-690

[10] Kumar K. Efficient Networks Communication Routing Using Swarm Intelligence. Punjab, India: Modern Education and Computer Science Press; 2012

[11] Maniezzo V, Gambardella LM, Luigi FD. Ant colony optimization. In: New Optimization Techniques in Engineering. Bangalore, India: Springer; 2004. pp. 101-121

[12] Parpinelli RS, Lopes HS, Freitas AA. Data mining with an ant colony optimization algorithm. IEEE Transactions on Evolutionary Computation. 2002;6(4):321-332

[13] Roopa C, Harish B, Kumar SA. Segmenting ECG and MRI data using ant colony optimization. International Journal of Artificial Intelligence and Soft Computing. 2019;7(1):46-58

[14] Roy S, Biswas S, Chaudhuri SS. Nature-inspired swarm intelligence and its applications. International Journal of Modern Education & Computer Science. 2014;6(12):55-65

[15] Tsai H-C. Integrating the artificial bee colony and bees algorithm to face constrained optimization problems. Information Sciences. 2014;258:80-93

Chapter 2

Application of Ant Colony Optimization for Co-Design of Hybrid Electric Vehicles

Majid Vafaeipour, Dai-Duong Tran, Thomas Geury,
Mohamed El Baghdadi and Omar Hegazy

Abstract

One key subject matter for effective use of Hybrid Electric Vehicles (HEVs) is searching for drivetrains which their component dimensions and control parameters are co-optimally designed for a desired performance. This makes the design challenge as a problem, which needs to be addressed in a holistic way meeting various constraints. Along this line, the strong coupling between components sizes of a drivetrain and parameters of its controllers turns the optimal sizing and control design of HEVs into a Bi-level optimization problem. In this chapter, an important application of continuous Ant Colony Optimization (ACO_R) for integrated sizing and control design of HEVs is thoroughly discussed for minimizing the drivetrain cost, minimizing the fuel consumption and addressing the control objectives at the meantime. The outcome of this chapter provides useful information related to incorporation of soft-computing, modeling and simulation concepts into optimization-based design of HEVs from all respects for designers and automotive engineers. It brings opportunities to the readers for understanding the criteria, constraints, and objective functions required for the optimal design of HEVs. Via introducing a two-folded iterative framework, fuel consumption and component sizing minimizations are of the main goals to be simultaneously addressed in this chapter using ACO_R.

Keywords: Hybrid Electric Vehicles, Continuous Ant Colony Optimization, Integrated Design, Modeling and Simulation, Parallel HEV, Energy Management Strategy

1. Introduction

With the advent of hybridization concepts into the automotive field, searching for drivetrains which their component dimensions and control parameters are simultaneously designed for optimal objectives has been attained huge attention from the researchers. The hybrid drivetrains comprise several energy sources and components such as electric motors, batteries, power electronics converters and Internal Combustion Engine (ICE). Hence, making concrete design decisions for their topologies is significantly complicated compared to conventional ones in terms of sizing. Furthermore, the design space becomes larger considering complexities caused by indispensable power control parameters and consequently high degrees of freedom due to presence of multiple power sources [1, 2]. This produces a large search space making it sophisticated for achieving objectives which are often counteracting, but equally important, e.g. satisfactory charge maintenance and fuel consumption minimization [3–6].

Due to their inevitable interrelations, the design levels of drivetrains cannot be performed independently or through standalone sequential framework as it leads one to suboptimal results. This makes the design challenge as a problem, which needs to be addressed in a holistic way meeting various constraints. Along this line, the strong coupling between components sizes of a drivetrain and parameters of its controllers turns the optimal sizing and control design of HEVs into a Bi-level optimization problem. For obtaining an optimal system design, the drivetrain components dimensions and the vehicle energy management strategy (EMS) should be designed in an interconnected and cohesive manner called **integrated optimal design or co-design** leading to minimum drivetrain cost and minimum fuel consumption as main objectives. There are several optimization algorithms and sequences available for integrated design of HEVs such as stochastic, gradient-based, deterministic, and derivative-free optimization methods [7]. The algorithm selection for integrated design of hybrid drivetrains depends on design targets. However, among variety of existing approaches, the metaheuristic algorithms e.g. Genetic Algorithm (GA), Particle Swarm Optimization (PSO), Simulated Annealing (SA) etc., owing to their derivative-free features, could bring great potential and flexibility toward handling the non-monotonic, non-linear, and highly dynamic nature of HEV design.

In this chapter, an application of continuous Ant Colony Optimization (ACO) as a relatively recent nature-inspired algorithm is presented for integrated design of HEVs focusing on minimization of drivetrain costs besides fuel consumption at the meantime. The design variables include power rating of the components (i.e. battery, ICE, electric motors) and control parameters dealing with power sharing through the components. Various equality and inequality constraints involve in the optimization procedure related to components power sharing limitations, initial and final battery state-of-charge (SoC), maximum and minimum allowable SoC boundaries, and charging rate limitations. To this end, first there is a need to establish a full vehicle model and its corresponding energy management strategy (EMS) which will be performed in Simulink® environment. A modeled passenger vehicle will be coupled into an ACO algorithm scripted in MATLAB to work in tandem for the optimization purpose. The developed framework triggers the integrated design objectives via minimizing sizing and control objective functions while satisfying the design constraints to be eventually compared with an initial non-optimal case. The optimization includes two iterative nested parts linked into each other through an inner loop to consider the optimization objective and constraints for component sizing and control in an integrated and iterative manner as simplified in **Figure 1**.

The present chapter is organized as follows. Section 2 presents the drivetrain architecture of the studied passenger HEV. In Section 3, individual modeling of the vehicle's components, EMS and corresponding descriptions will be elaborated. Section 4 reviews the principles of the used ACO algorithm. Section 5 narrows down

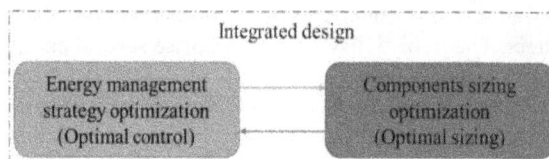

Figure 1.
Coordination of the nested integrated design.

the objective functions, optimization constraints and the integration of the simulation into optimization process for the studied application. Section 6 discusses and compares the attained results, and finally Section 7 recapitulates the conclusions. The outcome of this chapter provides useful information related to incorporation of soft-computing, modeling and simulation concepts into optimization-based design of HEVs from all respects for designers and automotive engineers.

2. Drivetrain architecture

In general, the HEVs are a combination of conventional and full-electric vehicles using both ICE and electric motor/generator for propulsion. Various topology of HEVs (e.g. series, parallel, series–parallel) exist depending on how the comprising power sources and components are connected through the vehicle structure. The parallel architecture for a passenger HEV is considered for the drivetrain topology of this study. The parallel drivetrain utilizes more than one direct power source in its architecture to provide energy for the propulsion system. The ICE and electric motor (EM) in such a topology can be coupled/decoupled to the wheels when required which brings more degree of freedoms (DoFs) of operating the vehicle in different modes. Hence, the traction force can be provided by means of both ICE and EM or either of them independently leading to lower number of energy conversions and consequently lower losses in such a topology compared to the series HEVs [8]. In a parallel drivetrain the wheels can receive the generated power from the EM plus the one received from ICE. Since the EM can operate as an electric generator in such a topology, the battery pack can be charged during regenerative braking or when the ICE output power is greater than the required power at the wheels. **Figure 2** illustrates a schematic of the considered parallel HEV architecture.

3. Modeling of the vehicle subsystems

Three main approaches exist for modeling and simulation of electric vehicles topologies:

1. the kinematic (backward-facing) approach,

2. the quasi static (forward-facing), and

3. the dynamic approaches.

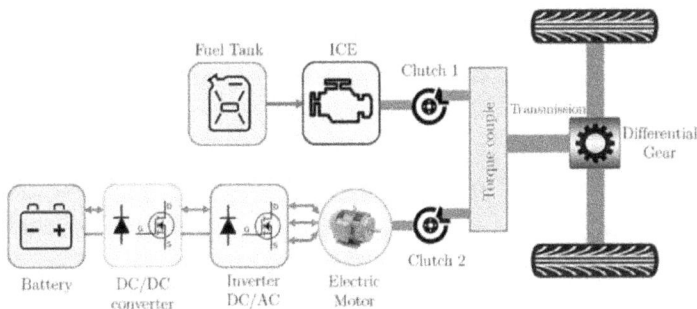

Figure 2.
Schematic of a parallel HEV topology.

The backward-facing and forward-facing approaches are also known as "effect-cause" and "cause-effect", respectively. Since a backward methodology carries out significant advantages such as simplicity and low computational cost in model-in--the-loop applications it is the most ideal testbed for integration into optimization algorithms requiring iterative operations [9]. Therefore, the backward-facing method is used for drivetrain modeling and simulation phase of this study. In principle, the backward facing calculation starts from the driving cycle velocity inputs to calculate the required tractive force at the wheel for propulsion. The required power, the translated torque and rotational speed will be calculated in a backward direction distributed through the components considering the power-split control block defined in an EMS subsystem. In this regard, **Figure 3** illustrates the calculation direction of a backward-facing model in a simplified way. The detailed modeling process of the subsystems are provided as follows.

The driving cycles are velocity time series representing a driving pattern; bring the road to a computer simulation and provide the profile that a vehicle requires to follow. The use of driving cycles assists modeling the drivetrain and the required performance to be considered for an appropriate design [8]. The standard New European Driving Cycle (NEDC), as represented in **Figure 4**, as the time dependent dynamic input of simulation process is used in this chapter.

A vehicle simulation model is required to be linked into the optimization algorithm for optimized integrated design and evaluation of the vehicle performance over the considered driving cycle. Hence, an energetic vehicle model based on the longitudinal dynamic motion laws is developed in MATLAB/Simulink® in this study. The vehicle longitudinal dynamic model uses speed and acceleration timeseries of a driving cycle to calculate the required tractive forces considering the

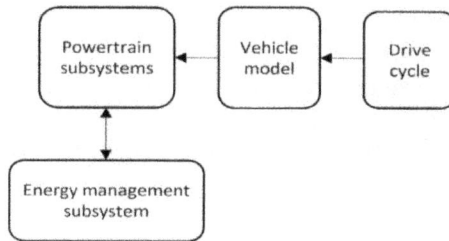

Figure 3.
Calculations direction in a generic backward-looking modeling.

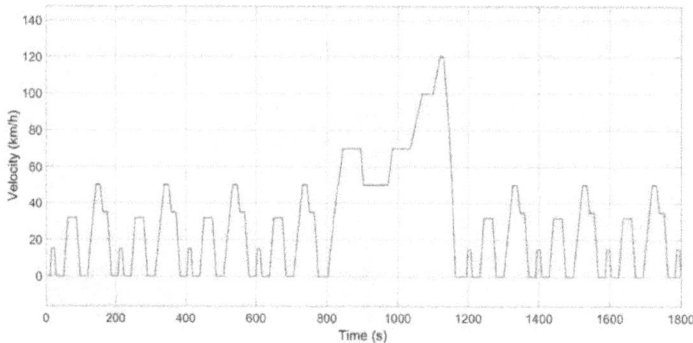

Figure 4.
Standard NEDC driving cycle, velocity profile [10].

drag resistance force, the rolling resistance force, the gradient resistance force, and the inertia force:

$$F_T = \frac{1}{2}\rho v^2 C_D A + C_r mg \cos a + mg \sin a + mC_J \frac{dv}{dt} \tag{1}$$

where its constant values are described and given in **Table 1**.

Consequently, the torque T_w and the rotational speed ω_w required to be supplied can be modeled. Along this line, by knowing the wheels radius R_w one can readily have the output of vehicle dynamic model to be fed into transmission subsystem model:

$$T_w = F_T R_w \tag{2}$$

$$\omega_w = \frac{v}{R_w} \tag{3}$$

In general, the vehicle components can be modeled using physical equations, analytical models (i.e. equivalent circuit) or considering related efficiency maps, which relate torque-speed or voltage–current pairs to their corresponding efficiency [11]. Using the obtained input torque and rotational speed values, the efficiency map defined in a look-up-table (LUT), power flow through the Electric Motor (EM) can mathematically be expressed as:

$$T_G = T_w G_r \eta^{\beta} \tag{4}$$

$$\omega_G = \omega_w G_r \tag{5}$$

It is notable that the efficiency term in Eqs. (4) and (5) must be treated contrarily for motoring and regenerating braking modes having positive and negative power flows, respectively. To this end, the efficiency operators $\beta = -1$ for the motoring mode (P > 0), and $\beta = 1$ for the braking mode (P < 0) are considered in the modeling process.

Figure 5 represents the efficiency map of the 75kw EM considered for the present study stored in EM LUT which can be scaled by torque and consequently power as an EM sizing decision variable in the optimization procedure.

$$P = T_{EM}\omega_{EM}\eta^{\beta}(T_{EM}, \omega_{EM}) \tag{6}$$

Description	Parameter (unit)	Quantity
Mass	m (kg)	1350
Drag coefficient	C_D	0.24
Rolling resistance coefficient	C_r	0.009
Rotational inertia coefficient	C_J	1.075
Frontal area	A (m²)	1.74
Wheel radius	R_w (m)	0.287
Air Density	ρ (kg/m³)	1.2
Gravitational acceleration	g (m/s²)	9.8
Road slope	a (degree)	0

Table 1.
Constants of vehicle dynamic calculation.

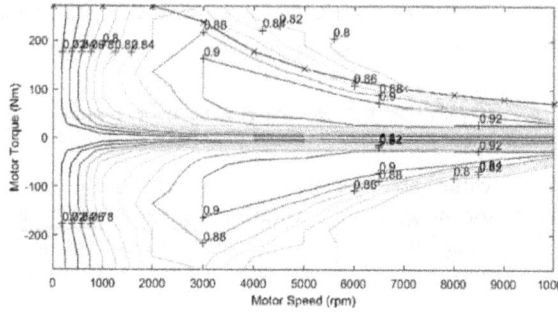

Figure 5.
75 kW EM efficiency map [12].

Similarly, the core functionality of the ICE subsystem used in this study is based on an input–output approach using torque-speeds pairs corresponded to the efficiency and fuel rate map stored into LUTs in the vehicle model. Having the output fuel consumption rates data and the fuel density, the consumed fuel in liter can be modeled in fuel tank subsystem as given in Eq. (7), where $\dot{m}$ represents the fuel consumption rate and ρ_f is the fuel density [13]. **Figure 6** represents the efficiency map of the 41 kW engine considered for the present study which can be scaled by torque and consequently power as a sizing variable in the optimization procedure.

$$Fuel = \int_0^t \frac{\dot{m}}{\rho_f} dt \tag{7}$$

A lithium-ion battery pack based on a semi-empirical first order Thevenin equivalent circuit is modeled in the battery subsystem. The elements of the battery model can be identified by using the experimental data [14] for open circuit voltage (V_{oc}), the internal resistance (R_{int}), the polarization capacitance (Cp), and the polarization resistance (R_p), which are stored in the LUTs of the corresponding subsystem. The terminal voltage of the pack V_{batt} and SoC can be expressed as:

$$I_{batt} = \frac{I_{load}}{N_{Batt}} \tag{8}$$

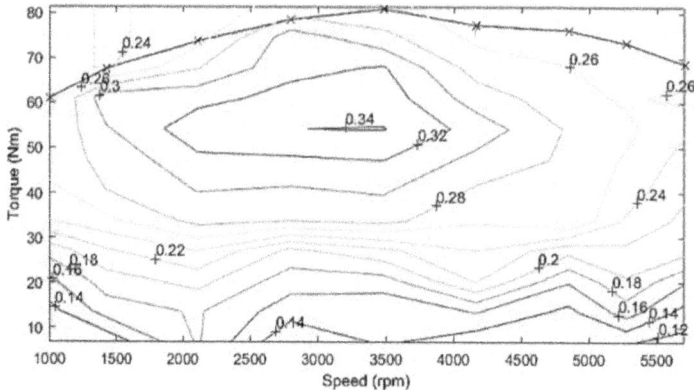

Figure 6.
41 kW ICE efficiency map [12].

$$\frac{dV_{cp}}{dt} = \frac{-V_{cp}}{C_p R_p} + \frac{I_{Batt}}{C_p} \qquad (9)$$

$$V_{Batt} = N_{Batts}\left(V_{oc} - I_{Batt}R_{int} - V_{cp}\right) \qquad (10)$$

$$SoC = SoC_0 + \frac{1}{3600}\int \frac{I_{Batt}}{C_b}dt \qquad (11)$$

Table 2 provides the specification of LiFePO$_4$ (LFP) battery cells used for modeling while number of cells are considered as the battery power sizing decision variable in the optimization procedure.

The output of power converters is modeled considering the power flow calculation direction and the components efficiency are used in their corresponding LUTs. The operators $\beta = -1$, and $\beta = 1$ are considered for the motoring mode (while P > 0), and the braking mode (while P < 0), respectively.

$$P_{out} = P_{in}\eta^{\beta} \qquad (12)$$

The main role of the energy management strategy (EMS) subsystem in HEVs is to define power sharing control principles satisfying set of required control objectives. The control strategies are mainly categorized into rule-based (RB) and optimization-based (OB) ones. The RB strategies as they are structurally working under If-Then rules, may handle trivial control objectives (e.g. HEV battery charge-sustaining), however, they are highly fragile in leading to optimal results when it comes to fuel consumption minimization. Hence, there is a need for coupling RB strategies into OB strategies to form a robust control framework as considered in the context of the present chapter. To this end, a RB strategy considering different vehicle operation modes is linked to a Low Pass Filter (LPF) OB strategy in the EMS block of the modeled vehicle to satisfy control optimization constraints and objectives. The RB control part updates the operating modes through the simulation considering the requested load, speed, accessible power from energy sources, battery state-of-charge (SoC) and power split control variables. The operating modes considering these objectives can be categorized as follows:

- Pure electric mode;

- Hybrid-traction mode;

- Engine traction and battery charging mode;

Parameter (unit)	Quantity
Rated capacity (Ah)	14
Nominal voltage (V)	3.6 V
Max discharging current (A)	100 A
SoC$_0$ (%)	80%
Min Voltage (V)	2.5
Max Voltage (V)	4.15
C_rate limit while charging	−3

Table 2.
LiFePO$_4$ battery cell parameters.

- Hybrid battery charging by both ICE and regenerative breaking;

- Regenerative braking mode.

However, the fuel consumption is significantly depended not only on the defined operating rules, but also on the OB power-split method used, specifically for the hybrid operating modes for improving efficiency and control robustness. Hence, an optimized Low Pass Filter (LPF) strategy will be introduced to the optimization algorithm to optimize the power-split control part by finding the best sharing control variable of LPF strategy satisfying power sharing objectives and constraints. In this regard, the considered OB-LPF strategy optimizes power sharing between the supplying components (i.e. battery and ICE) to provide required driving power while minimizing fuel consumption. Through using a filter-based transfer function, the filtered component of the required power passes to be supplied by the ICE while its difference with the total demand will be supplied by the battery subsystem [15]. The standard transfer function defined in the energy management subsystem and optimization process is considered as below. Here the LPF denominator (τ) is the control variable to be searched through optimization routine toward having the control objectives and constraints satisfied:

$$f_{LPF} = \frac{1}{\tau.s + 1} \tag{13}$$

The elaborated subsystems are integrated in the Simulink® environment to form the whole vehicle model to work in tandem with a MATLAB-based ANT Colony (ACO_R) algorithm for component sizing and control optimization.

4. ACO_R algorithm

The metaheuristic Ant Colony Optimization (ACO) system, inspired by foraging behavior of ants, was first developed by Dorigo et al. [16] for discrete optimization problems. In the discrete ACO the ants represent stochastic procedures toward establishing set of candidate solutions in presence of a pheromone model. The pheromone model encompasses numerical values as pheromones being updated in iterations leading ants to promising solution regions of the search space. Hence, in the discrete ACO, pheromone information is used in a sampling process to construct a discrete probability function based on the sorted solutions. Later on for solving continuous domains, Socha and Dorigo [17] developed the continuous ACO (ACO_R) which can utilize continuous multimodal probability functions such as weighted Gaussian functions over the search space to solve a non-linear function optimization problem as Min $f(x) : a \leq x \leq b$ where vector $x = (x1, ..., xn)$ represents the decision variables having vectors a and b as the lower and upper search space boundaries, respectively [18]. To this end, it produces a probability density function for each iteration using solution archives as an explicit memory of the search history in the pheromone model. Accordingly, the ACO_R used in this study includes three main phases as:

- Pheromone representation;

- Probabilistic solution construction;

- Pheromone update.

In this regard, first in pheromone representation stage the algorithm uniformly and in a random manner initializes the solution archive of k solutions where each solution is a D-dimensional vector for $x_i \in [x_{\min}, x_{\max}]$ where $i = 1, 2, \ldots, D$. The archived solutions are sorted based on their quality (best to worth). In the probabilistic construction stage a solution (i.e. S_j) for j_{th} solution will be selected considering the choosing probability p_j defined by a Gaussian probability function where each solution S_j is corresponded to its weight w_j, mathematically expressed as follows [19, 20]:

$$w_j = \frac{1}{qk\sqrt{2\pi}} \exp\left(\frac{-(rank(j)-1)^2}{2q^2k^2}\right) \tag{14}$$

$$p_j = \frac{w_j}{\sum_{a=1}^{k} w_a} \tag{15}$$

In this regard, the better solutions would get higher choosing chances. Correspondingly, $rank(j)$ is the rank of sorted solution S_j, and the intensification factor (selection pressure factor), q, is a modifiable algorithm parameter dealing with uniformity of the probability function while larger q values make the probability function more uniform. A solution would be chosen based on the probabilistic approach and new candidate solutions are generated as the algorithm samples neighborhood of i_{th} decision variable, S_{guide}^i, using the Gaussian function G (see **Figure 7**) with mean $\mu_{guide}^i = S_{guide}^i$ and standard deviation σ_{guide}^i values as follows [21]:

$$\sigma_{guide}^i = \xi \sum_{r=1}^{k} \frac{\left|S_r^i - S_{guide}^i\right|}{k-1} \tag{16}$$

It calculates the average distance value of the i_{th} component of S_{guide} and the values of the i_{th} components of solutions in the archive. Here the multiplier $\xi > 0$ is

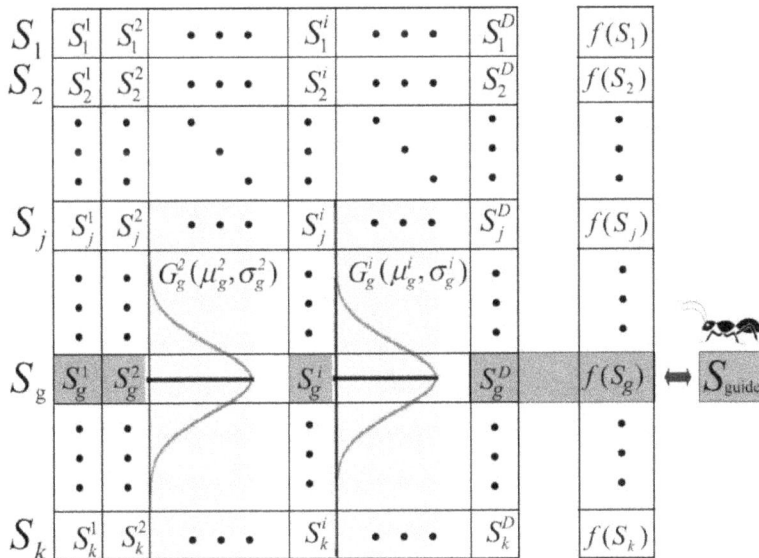

Figure 7.
The solution archive and the Gaussian functions used in ACO_R [21].

the user-specified pheromone evaporation rate parameter affecting the convergence while lower ξ values lead to lower convergence speed.

In the phoremone update phase, the process repeats for Na (number of ants) times while appending the new generated solutions to the k solutions of the archive, to incrementally sort $k + Na$ solutions and remove the worst solutions. Therefore, before a next iteration starts, the algorithm updates the archive keeping only the best k solution and discarding the worst ones having the archive size unchanged. For a considered number of itterations, the algorithm runs till reaching a stoping

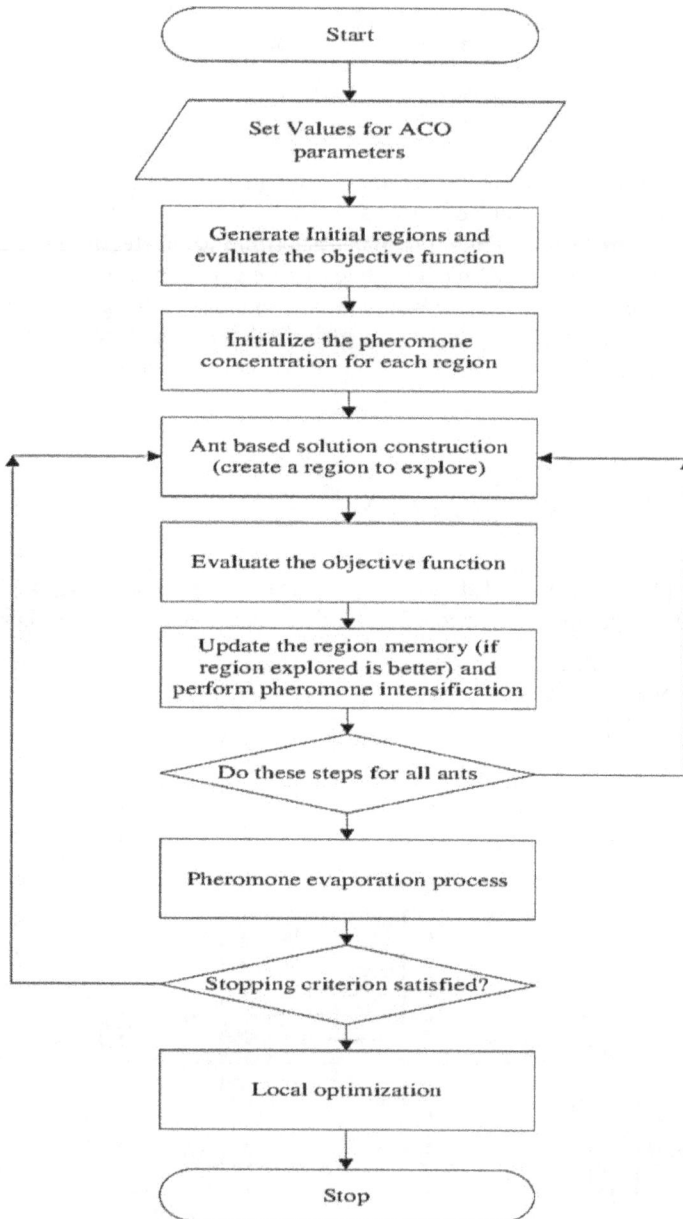

Figure 8.
General flowchart of the algorithm for the inner loops of the nested framework [22].

criteria to eventually select the best solution among the the evaluated positions. **Figure 8** provides the algorithm flowchart for the elaborated procedures.

5. Incorporation of the vehicle model into the optimization procedure, objectives and constraints

The established Simulink® model previously explained in this chapter is integrated to the MATLAB-based ACO script to iteratively work in tandem for the optimization purpose. The framework considers set of defined control and component sizing constraints and objectives. The co-design optimization process includes two iterative phases linked into each other through an outer loop to consider finding optimized EMS and component sizes at the meantime as simplified in **Figure 9**.

Two objective functions corresponding to the fuel consumption and components cost are considered for control and sizing optimizations, respectively. For the control parameter optimization, the decision variable would be the previously introduced LPF denominator (τ). Therefore, for the EMS optimization the algorithm aims to search for the power sharing variable which minimizes the fuel consumption (FC) while satisfying the constraints.

$$
\min\left(FC\right) \;=\; \min J_1 \;=\; \min \int_0^t \dot{m} dt \tag{17}
$$

On the other hand, another objective function is used for the component sizing formed based on the cost of powertrain components considering their prices per

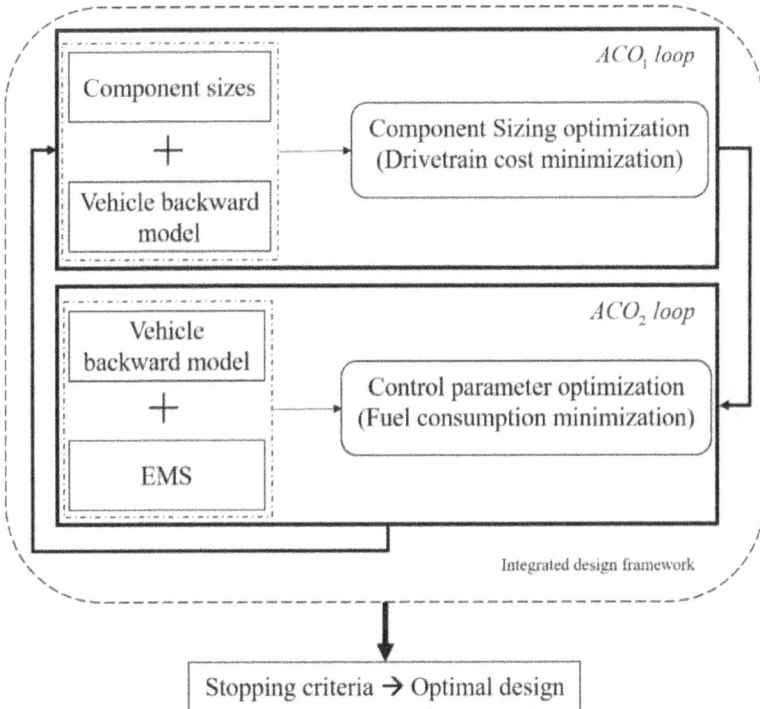

Figure 9.
Coordination architecture of the co-design and variables interrelations.

power unit. In other words, for the optimal sizing, the algorithm searches for the sizes which minimize the powertrain cost while satisfying the constraints. In this regard, the following formulation can be readily expressed for this objective function:

$$\min\left(Cost_{powertrain}\right) = \min J = \underset{sizes}{\arg\min}\left(\text{€}C_{ICE} + \text{€}C_{EM} + \text{€}C_{inv} + \text{€}C_{Batt} + \text{€}C_{conv}\right)$$

(18)

where the cost, €, for each component is considered in Euros and can be calculated based on per-power unit price, Q_{comp}, of each component considering its size:

$$\text{€}C_{comp} = \left(Q_{comp}\right)\left(size_{comp}\right)$$

(19)

The used per power unit price are given in **Table 3** for the ICE, the battery and the DC-DC converter while the inverter cost can be directly included based on the following equation.

$$\text{€}C_{inv} = 13.26(P)^{1.1718}$$

(20)

For minimization of the objective functions, the charge sustaining HEV is subjected to the following inequality constrains:

$$|SoC_f - SoC_i| < \varepsilon_0$$

(21)

$$SoC_{min} - \varepsilon < SoC(t) < SoC_{max} + \varepsilon$$

(22)

$$C\_Rate(t) \geq -3; Negative\ sign\ stands\ for\ charging$$

(23)

where the sizes of components are bounded between the considered minimum and maximum values of the search space. Regarding the SoC, constraint in Eq. (21) indicates the charge sustaining requirement, and Eq. (22) stands for the allowable limits of the SoC over the total driving cycle. The constraint in Eq. (23) is considered based on LiFePO$_4$ battery type chemistry to avoid sudden charges, to avoid fast aging of the battery pack, and to improve battery's lifetime and performance. It is notable that some constraints must be incorporated into the objective function as penalties to penalize the cost via adding (in minimization problems) or deducting (in maximization problems) a big enough penalty value when the constraint(s) is violated. This technique is useful to consider the inequality constraints which cannot be directly involved in the formulations of the objective function. As the optimization problem for both objectives are both minimization type here, the added penalty is considered.

Component	Q	Unit
Q_{ICE}	80	€/kW
Q_{Batt}	200	€/kWh
Q_{DC-DC}	100	€/kW
Q_{EM}	90	€/kW

Table 3.
Per-power unit prices used in cost objective function.

6. Results and discussion

To investigate the effectiveness of the proposed framework, simulation and optimization over NEDC driving cycle are performed and the results are provided in this section. The comparisons are considered for an initial non-optimized case (before integrated design) versus optimized cases (after integrated design). The main objective of the integrated design is to minimize component sizes and as a result the cost of powertrain besides achieving optimized fuel consumption while satisfying the constraints through the developed nested iterative framework. For achieving close enough values of the initial and final SoC, $\varepsilon_0 = 0.3\%$, and for providing slight degree of freedom on allowable SoC_{min} and SoC_{max}, small ε allowable sliding value as 4%, were all considered in the formulations of the optimization constraints. **Figure 10** presents the power sharing between the battery and the ICE satisfying driving power. In addition, evolution of the battery SoC and C-Rate for the studied driving cycle after the integrated design are plotted in the same figure. As can be seen, the regulated EMS could successfully recover the SoC to achieve close values for initial and final SoC over the full cycle ($SoC_f \simeq SoC_f$) having the ICE charging the battery when needed while considering the defined $C\_Rate(t)$ violation limit at the meantime. In addition, the SoC allowable minimum and maximum boundary is satisfied through the desired window range for the whole cycle. Consequently, **Table 4** provides detailed evaluations in terms of control constraints satisfaction related to triggered EMS goals.

Correspondingly, **Table 5** summarizes the design parameters before and after optimal integrated design while fuel consumption besides powertrain cost

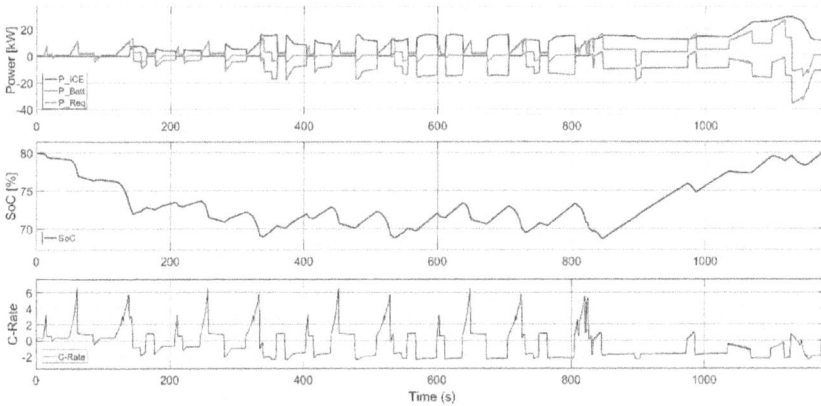

Figure 10.
Power distribution (kW), SoC (%) recovery, and C-rate results.

Considered features	Before	After
$\|SoC_f - SoC_i\| < 0.3$	✗	✓
$SoC_{min} - \varepsilon < SoC(t) < SoC_{max} + \varepsilon$	✗	✓
Driving power needs	✓	✓
$C\_Rate(t) < -3$	✓	✓
All EMS objectives satisfied?	✗	

Table 4.
Control goals satisfaction.

Design variable	Description	Lower bound	Upper bound	Initial value	Optimal design value
P_{ICE} (kW)	ICE size	30	120	84	75
Cap_{Batt} (kWh)	Battery pack size	3	20	9	7
EM (kW)	Electric motor size	50	120	97	80

Table 5.
Component sizes before and after integrated design.

Objectives	Before integrated design	After integrated design
Fuel Consumption (L/100 km)	5.1	4.8
Improvement (%)	—	5
Powertrain Cost (Euros)	28100	23200
Improvement (%)	—	17

Table 6.
Fuel consumption and powertrain cost improvements.

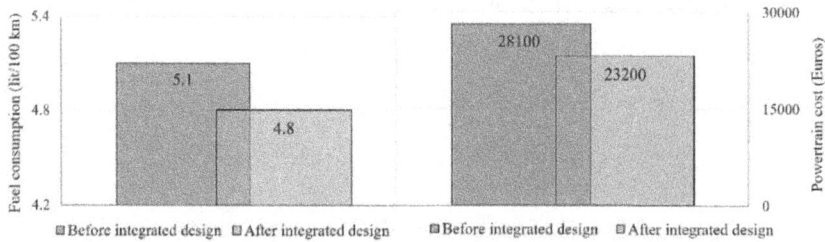

Figure 11.
Fuel consumption and powertrain cost comparisons.

improvements for the studied cases are provided in **Table 6**. The sizes of the decision variable values attained after integrated design indicates that the algorithm could efficiently downsize the component sizes and consequently the powertrain costs. It can be observed that improvements are achieved in the fuel consumption and cost of powertrain components after the co-design design by 5% and 17%, respectively as illustrated in **Figure 11**.

7. Chapter conclusions

This chapter investigated a combination of optimization-based and rule-based energy management strategies to perform an integrated design approach for a passenger hybrid electric vehicle use-case. The modeling procedure of the components were presented, and the corresponding Simulink® model was developed and linked to an ACO_R algorithm to work iteratively for the co-design design purpose. To check the performance of the proposed framework, simulations and optimizations were carried out over the NEDC driving cycle. The detailed results through the cycle for power splitting, battery SoC values, battery C_{rate} values, fuel consumption and powertrain costs were obtained and compared for before and after applying the approach. The results indicated that the proposed framework not only was able to provide an acceptable management regarding the battery SoC and C_{rate}, but also was competent of bringing significant added values in terms of the fuel consumption and powertrain cost reduction. The outcome of the present study paves the path for experimental Hardware-in-the-Loop and Vehicle-in-the-Loop validations.

Acknowledgements

The authors are grateful to Flanders Make (FM) for supporting our research group in the current work.

Conflict of interest

The authors declare no conflict of interest.

Author details

Majid Vafaeipour[1,2*], Dai-Duong Tran[1,2], Thomas Geury[1,2],
Mohamed El Baghdadi[1,2] and Omar Hegazy[1,2*]

1 ETEC Department and MOBI Research Group, Vrije Universiteit Brussel (VUB),
Brussel, Belgium

2 Flanders Make, Heverlee, Belgium

*Address all correspondence to: majid.vafaeipour@vub.be and
omar.hegazy@vub.be

IntechOpen

References

[1] Vafaeipour M, El Baghdadi M, Verbelen F, Sergeant P, Van Mierlo J, Hegazy O. Experimental implementation of power-split control strategies in a versatile hardware-in-the-loop laboratory test bench for hybrid electric vehicles equipped with electrical variable transmission. Applied Sciences 2020;10: 4253.

[2] Verbelen F, Lhomme W, Vinot E, Stuyts J, Vafaeipour M, Hegazy O, et al. Comparison of an optimized electrical variable transmission with the Toyota Hybrid System. Applied Energy 2020; 278:115616.

[3] Neffati A, Caux S, Fadel M. Fuzzy switching of fuzzy rules for energy management in HEV. IFAC Proceedings Volumes 2012;45:663.

[4] Wei Z, Xu J, Halim D. HEV power management control strategy for urban driving. Applied Energy 2017;194:705.

[5] Wei Z, Xu Z, Halim D. Study of HEV power management control strategy based on driving pattern recognition. Energy Procedia 2016;88:847.

[6] Wu J, Peng J, He H, Luo J. Comparative analysis on the rule-based control strategy of two typical hybrid electric vehicle powertrain. Energy Procedia 2016;104:384.

[7] Tran D-D, Vafaeipour M, El Baghdadi M, Barrero R, Van Mierlo J, Hegazy O. Thorough state-of-the-art analysis of electric and hybrid vehicle powertrains: Topologies and integrated energy management strategies. Renewable and Sustainable Energy Reviews 2020;119:109596.

[8] Vafaeipour M, El Baghdadi M, Verbelen F, Sergeant P, Van Mierlo J, Stockman K, et al. Technical assessment of utilizing an electrical variable transmission system in hybrid electric vehicles. 2018

IEEE Transportation Electrification Conference and Expo, Asia-Pacific (ITEC Asia-Pacific): IEEE; 2018, p. 1.

[9] Millo F, Rolando L, Andreata M. Numerical simulation for vehicle powertrain development. Numerical Analysis-Theory and Application: IntechOpen; 2011.

[10] www.unece.org. Accesible 2021.

[11] Vafaeipour M, El Baghdadi M, Van Mierlo J, Hegazy O, Verbelen F, Sergeant P. An ECMS-based approach for energy management of a HEV equipped with an electrical variable transmission. 2019 Fourteenth International Conference on Ecological Vehicles and Renewable Energies (EVER): IEEE; 2019, p. 1.

[12] Advisor, NREL. Accesible 2021.

[13] Vafaeipour M, Tran D-D, El Baghdadi M, Verbelen F, Sergeant P, Stockman K, et al. Optimized energy management strategy for a HEV equipped with an electrical variable transmission system. 32nd Electric Vehicle Symposium (EVS32); 2019.

[14] Hegazy O, Barrero R, Van Mierlo J, Lataire P, Omar N, Coosemans T. An advanced power electronics interface for electric vehicles applications. IEEE transactions on power electronics 2013; 28:5508.

[15] Vafaeipour M, El Baghdadi M, Tran D-D, Van Mierlo J, Hegazy O, Verbelen F, et al. Energy Management Strategy Optimization for Application of an Electrical Variable Transmission System in a Hybrid Electric City Bus. 2020 Fifteenth International Conference on Ecological Vehicles and Renewable Energies (EVER): IEEE; 2020, p. 1.

[16] Dorigo M, Maniezzo V, Colorni A. Ant system: optimization by a colony of

cooperating agents. IEEE Transactions on Systems, Man, and Cybernetics, Part B (Cybernetics) 1996;26:29.

[17] Socha K, Dorigo M. Ant colony optimization for continuous domains. European journal of operational research 2008;185:1155.

[18] Mathur M, Karale SB, Priye S, Jayaraman V, Kulkarni B. Ant colony approach to continuous function optimization. Industrial & engineering chemistry research 2000;39:3814.

[19] Blum C. Ant colony optimization: Introduction and recent trends. Physics of Life Reviews 2005;2:353.

[20] Omran MGH, Al-Sharhan S. Improved continuous Ant Colony Optimization algorithms for real-world engineering optimization problems. Engineering Applications of Artificial Intelligence 2019;85:818.

[21] Liao T, Stützle T, Montes de Oca MA, Dorigo M. A unified ant colony optimization algorithm for continuous optimization. European Journal of Operational Research 2014;234:597.

[22] Khanna A, Mishra A, Tiwari V, Gupta P. A literature-based survey on swarm intelligence inspired optimization technique. J Adv Technol Eng Sci 2015;3:452.

Chapter 3

Grid Map Merging with Ant Colony Optimization for Multi-Robot Systems

Heoncheol Lee

Abstract

Multi-robot systems have recently been in the spotlight in terms of efficiency in performing tasks. However, if there is no map in the working environment, each robot must perform SLAM which simultaneously performs localization and mapping the surrounding environments. To operate the multi-robot systems efficiently, the individual maps should be accurately merged into a collective map. If the initial correspondences among the robots are unknown or uncertain, the map merging task becomes challenging. This chapter presents a new approach to accurately conducting grid map merging with the Ant Colony Optimization (ACO) which is one of the well-known sampling-based optimization algorithms. The presented method was tested with one of the existing grid map merging algorithms and showed that the accuracy of grid map merging was improved by the ACO.

Keywords: Ant Colony Optimization, Intelligent Robot, Grid Map Merging, SLAM, Multi-Robot Systems

1. Introduction

Multi-robot systems [1] have recently been in the spotlight because of the advantage that it can perform a given task more efficiently than a single robot system and can perform several tasks at the same time. For the design and construction of such a multi-robot system, various algorithms which are not required in a single robot system are required. If a multi-robot system is operated in unknown environments, it needs to conduct multi-robot simultaneous localization and mapping (SLAM) [2] to acquire the poses of multiple robots and a collective map for operating the give task cooperatively without collisions. An example of a multi-robot system for multi-robot SLAM in unknown environments is shown in **Figure 1**. The cooperation module which conducts global multiple path planning, relative robot pose estimation, and multiple map merging can be placed on the leader robot or a central control system. The wireless router can be located in the leader robot or another place to cover the operation area of multiple robots. The bandwidth for the wireless communication depends on the size of the operation area and the map representation method. To conduct SLAM, each robot needs sensors to acquire environmental data. Based on the SLAM result, each robot can plan a local path and move toward its own goal safely.

The most frequently used sensor for SLAM is a light detection and ranging (LiDAR) [3] which measures ranges by targeting an object with a laser and

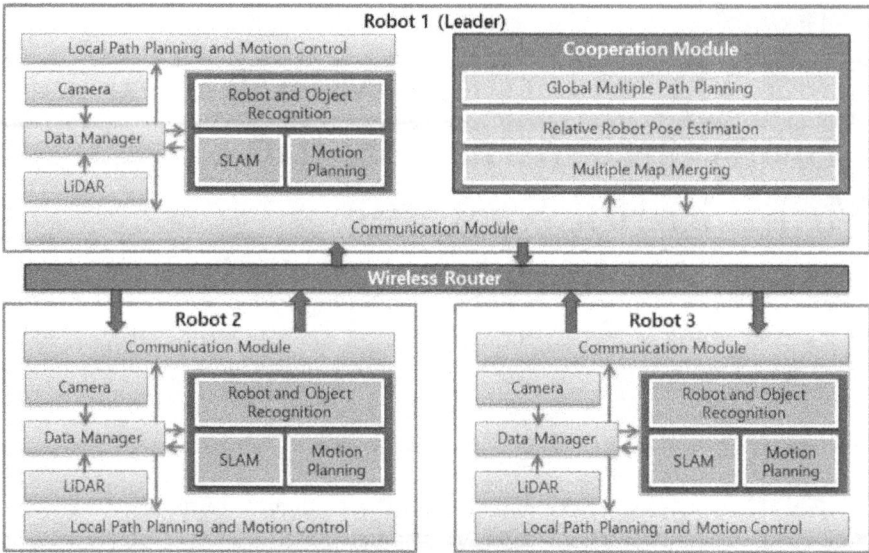

Figure 1.
An example of a multi-robot system in unknown environments.

measuring the time for the reflected light to return to the receiver. LiDAR can also be used to make digital three-dimensional representations of areas on the earth's surface and ocean bottom, due to differences in laser return times, and by varying laser wavelengths. Because a LiDAR can provide a lot of information about the surrounding environment, it has been used widely for SLAM. An example of using a LiDAR for a mobile robot is as shown in **Figure 2(a)**. If SLAM is conducted with a LiDAR, a map is generally represented by an occupancy grid map as shown in **Figure 2(b)**. The white, black and gray grids represent empty, occupied and unknown areas, respectively. The size of grids can be adjusted according to the resolution of the LiDAR and the memory size in the embedded system.

The key algorithm in multi-robot SLAM is the grid map merging algorithm in the cooperation module in **Figure 1** which accurately aligns and fuses the individual

Figure 2.
Occupancy grid map built by a mobile robot with a LiDAR sensor. (a) Mobile robot with a LiDAR sensor (b) Occupancy grid map.

grid maps of multiple robots. Many grid map merging algorithms have been developed, and they have their own advantages over others. However, for the more accurate grid map merging, all the algorithms need an optimization method to align the individual grid maps more precisely. In this work, we propose a new approach based on a sampling-based optimization method for grid map merging. The proposed approach was successfully conducted with other grid map matching algorithms and updated the map transformation matrix between robots more accurately.

The remainder of this paper is organized as follows. In Section 2, multi-robot SLAM is briefly described. In Section 3, the definition and classification of grid map merging are described. In Section 4, the proposed approach which is a grid map merging with ACO is presented. Section 5 shows and analyzes the experimental results of the proposed approach. Finally, conclusions are given.

2. Multi-robot SLAM

SLAM is to concurrently conduct two processes which are called localization and mapping, respectively. Mapping is to acquire a map of its surrounding environments to plan a path to its own goal without collisions with structures. Localization is to estimate its own pose within the acquired map. Unfortunately, SLAM is not easy because the two processes in SLAM depend on each other. In other words, the localization process requires a map as a reference to estimate its own pose, and the mapping process requires a pose which consists of location and orientation as a reference point to represent a map. Many researches have been conducted to conduct SLAM efficiently, and several nice solutions have been recently proposed. However, SLAM is still an open problem in the context of accuracy, reliability, and computational cost.

Multi-robot SLAM is to conduct the SLAM task using multiple robots for the sake of completing localization and mapping more efficiently. An example of configuring a two-robot SLAM is shown in **Figure 3**. Each robot conducts SLAM with its own sensors. Based on the multiple SLAM results gathered through the communication modules, the global state has been updated.

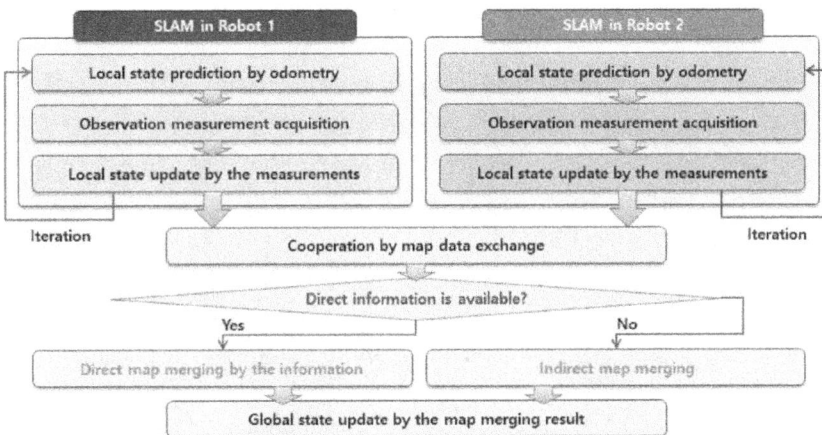

Figure 3.
An example of configuring a two-robot SLAM.

Due to the errors in sensors for multi-robot SLAM, the global state estimation is generally conducted with probabilistic formulations. The estimation of the two-robot SLAM state in **Figure 3** can be formulated as follows:

$$
\begin{aligned}
&P\left(x_{1:t}^1, x_{1:t}^2, M \mid z_{1:t}^1, u_{0:t-1}^1, x_0^1, z_{s:t}^2, u_{s:t-1}^2, \Delta_s^{21}\right) \\
&= P\left(M \mid x_{1:t}^1, z_{1:t}^1, x_{s:t}^2, z_{s:t}^2\right) P\left(x_{1:t}^1 \mid z_{1:t}^1, u_{0:t-1}^1, x_0^1\right) P\left(x_{s:t}^2 \mid z_{s:t}^2, u_{s:t-1}^2, x_s^1, \Delta_s^{21}\right)
\end{aligned}
\tag{1}
$$

where $x_{k:t}^i$ is the trajectory for robot i at times $k, k+1, \cdots, t$, and M is the merged map, and $u_{k-1:t-1}^i$ is the sequence of actions executed by robot i, and $z_{k:t}^i$ is the sequence of observations from robot i, and Δ_s^{21} is the relative pose between two robots at time s. Extended Kalman filters (EKF) [4] and Rao-Blackwellized particle filters (RBPF) [5] have been widely used as estimation methods for the probabilistic formulation. At the beginning of the estimation, the uncertainty of the state is large. But, as time goes, the uncertainty of the state has been gradually reduced if the observation measurements are acquired consistently, and data association is conducted properly. Especially, whenever loop closures [6] are conducted, the uncertainty of the state can be significantly reduced.

3. Grid map merging

The key algorithm to ensure the performance of multi-robot SLAM with LiDAR sensors is the grid map merging algorithm because even if the performance of the SLAM results of individual robots are good, the performance of multi-robot SLAM depends on the quality of the map transformation between robots. The concept of the grid map merging in multi-robot SLAM with LiDAR sensors is shown in **Figure 4**. Quantitatively, the grid map merging can be performed by acquiring a map transformation matrix T (MTM) which consists of translation amounts and a rotation angle between robots as follows:

$$
T\left(\Delta_x, \Delta_y, \Delta_\theta\right) =
\begin{bmatrix}
\cos \Delta_\theta & -\sin \Delta_\theta & \Delta_x \\
\sin \Delta_\theta & \cos \Delta_\theta & \Delta_y \\
0 & 0 & 1
\end{bmatrix}
\tag{2}
$$

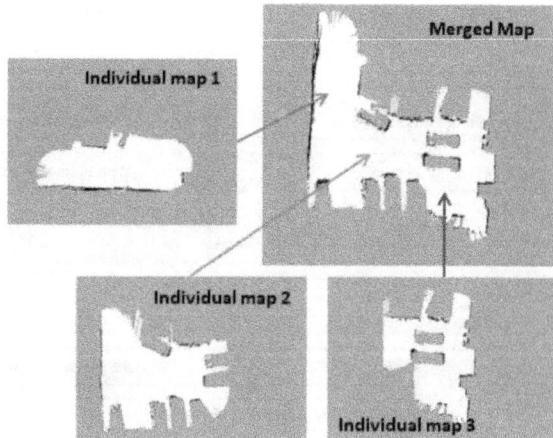

Figure 4.
The concept of the grid map merging in multi-robot SLAM with LiDAR sensors.

where Δ_x, Δ_y and Δ_θ are the translation amounts and a rotation angle between robots, respectively.

The method to find the MTM can be categorized into direct map merging and indirect map merging according to the existence of the direct sensor measurements between robots or common objects. The direct map merging is to directly acquire the map transformation matrix by obtaining the inter-robot measurements which consist of relative distance and orientation between robots, which can be performed under a rendezvous. The indirect map merging acquires the map transformation matrix by finding and matching the overlapping areas of the individual maps of robots, which is called map matching. The detailed categorization of them and the brief descriptions of the previous works are summarized in [7, 8]. They have their own advantages, but they require commonly an optimization method to update the MTM more accurately regardless of the type of map merging.

4. Ant colony optimization for grid map merging

Given an MTM T, the objective function Φ to evaluate how two individual maps M_1 and M_2 are well overlapped for the merged map optimization can be defined as follows:

$$\Phi(M_1, M_2, T) = \sum_{x=a_1}^{a_2} \sum_{y=b_1}^{b_2} M_1(x,y) \cdot [T \, M_2(x,y)] \qquad (3)$$

where $a_1 \leq x \leq a_2$ and $b_1 \leq y \leq b_2$ are the whole ranges of the x and y coordinates of M_1 and M_2. Because T includes sinusoidal functions for map rotation, the objective function Φ has nonlinearity and thus is hard to be solved in a closed form.

Therefore, the optimization of Φ for grid map merging needs to be considered with sampling-based optimization such as MCO (Monte-Carlo Optimization) [9], PSO (Particle Swarm Optimization) [10] and ACO (Ant-Colony Optimization) [11]. They require commonly much computation due to their own iterative property. Instead, they are easy to implement regardless of the complexity or nonlinearity of the objective function. Thus, it is a reasonable approach to apply sampling-based optimization methods to the merged map optimization. This paper applies the ACO to the merged map optimization because the ACO requires the relatively smaller number of samples than the MCO and the PSO in the case of the merged map optimization. The ACO is a probabilistic technique for solving computational problems which can be reduced to finding good paths through graphs. Artificial ants locate optimal solutions by moving through a parameter space representing all possible solutions. Real ants lay down pheromones directing each other to resources while exploring their environment. The simulated ants similarly record their positions and the quality of their solutions, so that in later simulation iterations more ants locate better solutions [12].

The ACO needs to be modified to be applied to the merged map optimization. Because an even slight variation in the rotation angle causes a largely different map merging result in grid map merging, the concept of pheromones in the ACO cannot be properly applied to finding the optimal rotation angle. Therefore, each sample in a search space consists of x and y translations except for a rotation angle. Besides, since the search space for x and y translations may be largely different, the search space for the ACO for grid map merging needs to be divided into two areas which contains the possible configurations of x and y translations respectively as shown in **Figure 5**.

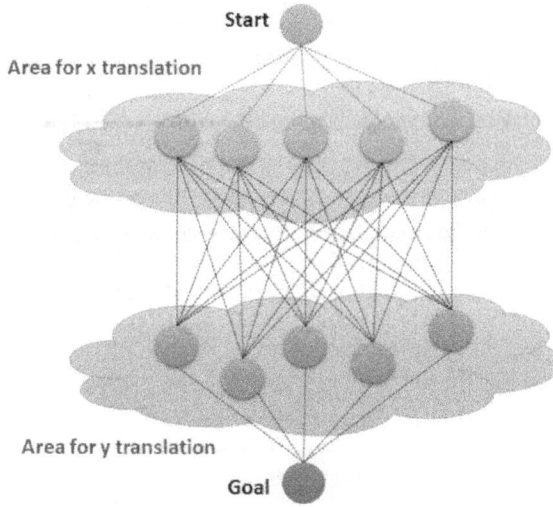

Figure 5.
The modified search space for the ACO for grid map merging.

In general, the i-th ant moves from state q to r with probability as follows:

$$p_{qr}^i = \frac{\left(\tau_{qr}\right)^\alpha \left(\eta_{qr}\right)^\beta}{\sum_{z \in allowed\ q} \left(\tau_{qz}\right)^\alpha \left(\eta_{qz}\right)^\beta} \tag{4}$$

where τ_{qr} is the amount of pheromone deposited for transition from state q to r. $0 \le \alpha$ is a parameter to control the influence of τ_{qr}, which was set to 1 in this work. η_{qr} is the desirability of state transition qr, which is typically set to the reciprocal value of the distance. $1 \le \beta$ is a parameter to control the influence of η_{qr}. τ_{qz} and η_{qz} represent the trail level and attractiveness for the other possible state transitions.

In the original ACO, the distance is the Euclidean distance between states. But, it needs to be redefined for grid map merging. In other words, the distance is not the Euclidean distance between the nodes but a new metric to evaluate how two individual grid maps are well overlapped. For a candidate tour of the i-th ant, $\Lambda_i = \left\{q_j^i, r_k^i\right\}$ where q_j^i and r_k^i are respectively the j-th and the k-th sample in the areas for x and y translations, the new metric Ψ is defined similarly to Eq. (3) as follows:

$$\Psi(\Lambda_i) = \frac{1}{\sum_{x=\tilde{a}_1}^{\tilde{a}_2} \sum_{y=\tilde{b}_1}^{\tilde{b}_2} M_1(x,y) \cdot \left[T\left(q_j^i, r_k^i, 0\right)\tilde{M}_2(x,y)\right]} \tag{5}$$

where $\tilde{M}_2$ is the transformed M_2 by a direct or indirect grid map merging algorithm. $\tilde{a}_1 \le x \le \tilde{a}_2$ and $\tilde{b}_1 \le y \le \tilde{b}_2$ are the whole ranges of the x and y coordinates of M_1 and $\tilde{M}_2$ after conducting the grid map merging algorithm. In this work, since the rotation angle is not a target of the merged map optimization with the ACO, the rotation angle in T is set to 0.

The global pheromone is updated as follows:

$$\tau_{qr} \leftarrow (1 - \rho)\tau_{qr} + \sum_{i}^{N_{ant}} \Delta\tau_{qr}^{i} \qquad (6)$$

where τ_{qr} is the amount of pheromone deposited for a state transition qr. ρ is the pheromone evaporation coefficient. N_{ant} is the number of ants. $\Delta\tau_{qr}^{i}$ is the amount of pheromone deposited by the i-th ant, which was set to $1/\Psi(\Lambda_i)$.

5. Experimental results

Before applying the proposed ACO to grid map merging, the spectra-based map merging (SMM) [13] algorithm was applied to find a coarse MTM. The SMM is a well-known indirect grid map merging algorithm which extracts spectral information from grid maps by the Hough transform and finds an MTM by matching the spectral information based on the cross-correlations. The individual grid maps in a multi-robot system were as shown in **Figure 6**. To reduce the computation time, each grid map was represented by a binary image with occupied (white) and unoccupied (black) grids.

Firstly, the rotation angle was coarsely estimated by the SMM. The Hough spectra and the cross-correlation between them are shown in **Figure 7**. The SMM estimates the rotation angle by taking the angle corresponding the maximum cross-correlation value. After rotating one of the individual grid maps by the estimated rotation angle, the SMM estimates the x and y translation amounts by taking the amounts corresponding the maximum x and y cross-correlation value. The x spectra and the x cross-correlations between them are shown in the top of **Figure 8**. Similarly, the y spectra and the y cross-correlations between them are shown in the bottom of **Figure 8**. The merged map by the rotation angle and the translation amounts estimated by the SMM is shown in **Figure 9**. The two individual grid maps were properly merged. But, they needs to be merged more accurately.

The proposed ACO for grid map merging was implemented based on an open source [14]. The settings for the ACO for grid map merging were as follow. The number of iterations was set to 50. The number of samples was set to 30. The number of ants N_{ant} was set to 100. The graphical results of the ACO for grid map merging are shown in **Figure 10**, which indicates that the pheromones were properly updated as time goes and found the optimal configuration of x and y translation amounts. In other word, the proposed method was successfully conducted and found the best x and y translation amounts. By the best x and y translation amounts and the rotation angle estimated by the SMM, the two individual grid maps were

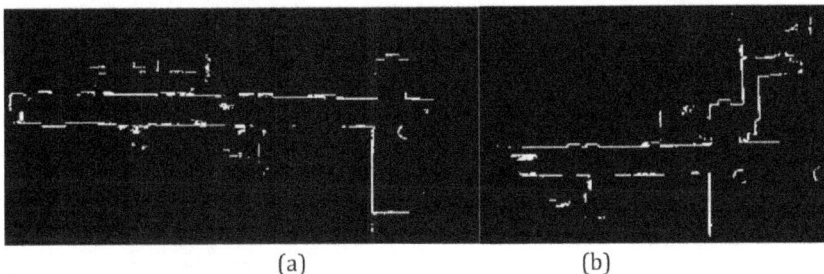

(a)   (b)

Figure 6.
Individual grid maps in a multi-robot system. (a) Individual grid map 1, $\mathbf{M}_1$ (b) Individual grid map 2, $\mathbf{M}_2$.

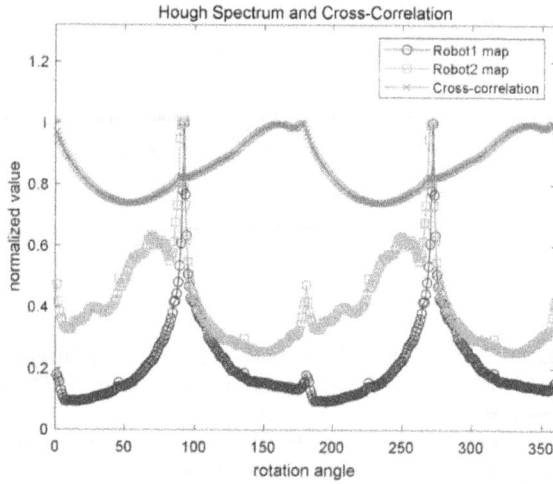

Figure 7.
Rotation angle estimation by the SMM.

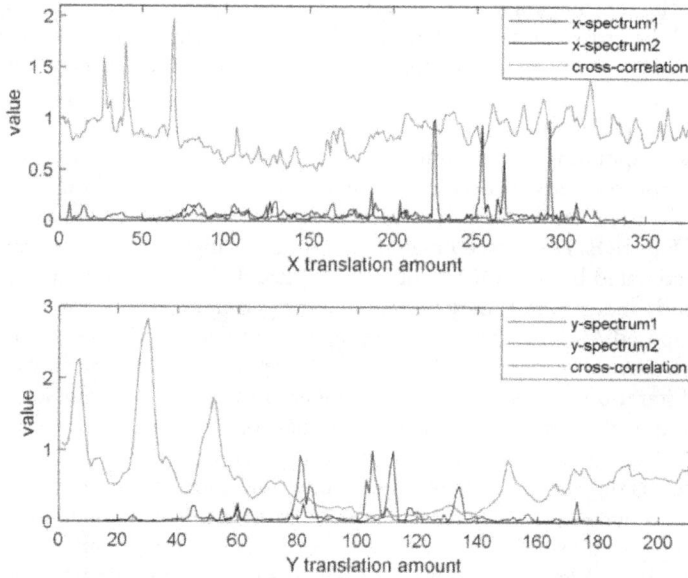

Figure 8.
Translation amounts estimation by the SMM.

merged more accurately as shown in **Figure 11**. Comparing with **Figure 9**, we can say that the error in the merged grid map was reduced.

The quantitative evaluation of the accuracy of grid map merging can be conducted with the following measure:

$$\text{Accuracy index} = \frac{\sum_{x=\hat{a}_1}^{\hat{a}_2} \sum_{y=\hat{b}_1}^{\hat{b}_2} M_1(x,y) \cdot \hat{M}_2(x,y)}{N_{overlap}} \qquad (7)$$

where $N_{overlap}$ is the number of commonly occupied grids in the overlapped areas when two individual grid maps are maximally overlapped, which is a global true

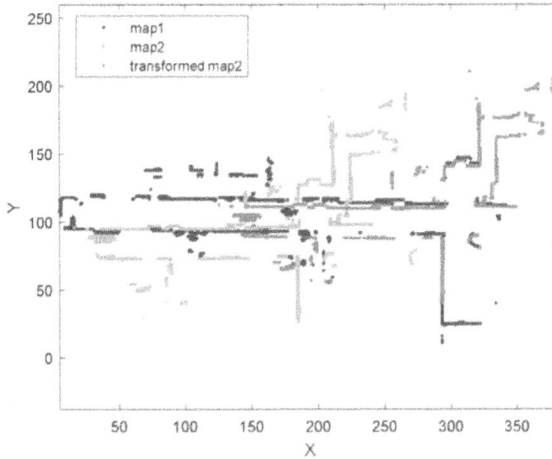

Figure 9.
The merged map by the SMM. The map 2 (green) was transformed by the SMM, and the transformed map 2 (red) was properly merged into map 1 (blue). However, they need to be merged more accurately.

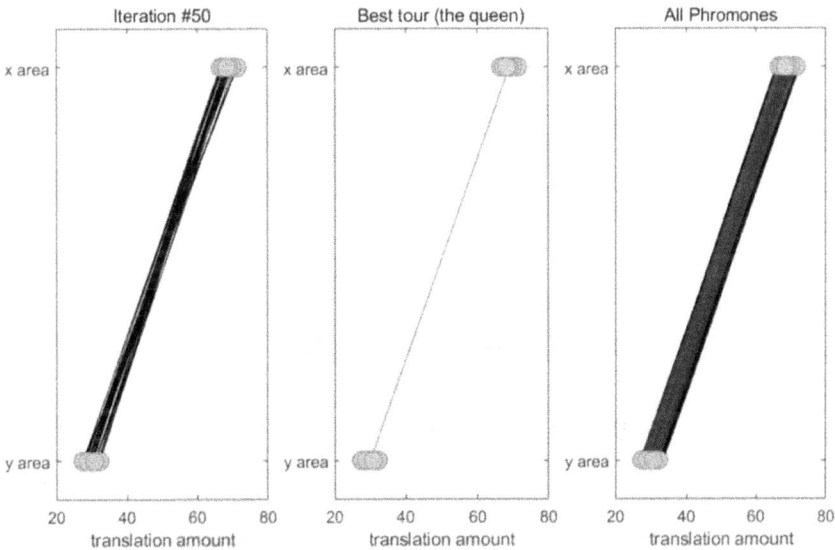

Figure 10.
ACO results for grid map merging. The red circles represent states in x and y areas. The left image represents the whole tours at each iteration. The middle image represents the best tour (the queen). The right image represents the pheromones along the tours.

value and not given to robots. $\hat{M}_2$ is the transformed M_2 by the ACO. $\hat{a}_1 \leq x \leq \hat{a}_2$ and $\hat{b}_1 \leq y \leq \hat{b}_2$ are the whole ranges of the x and y coordinates of M_1 and $\hat{M}_2$.

The map merging results of the proposed grid map merging method which uses both the SMM and the ACO was quantitatively compared with those of the only SMM-based grid map merging as shown in **Figure 12**. Because the performance of the ACO depends on the number of ants N_{ant}, the accuracy indices of the proposed method were analyzed with various N_{ant}. As expected, the ACO improved the accuracy of grid map merging with the SMM.

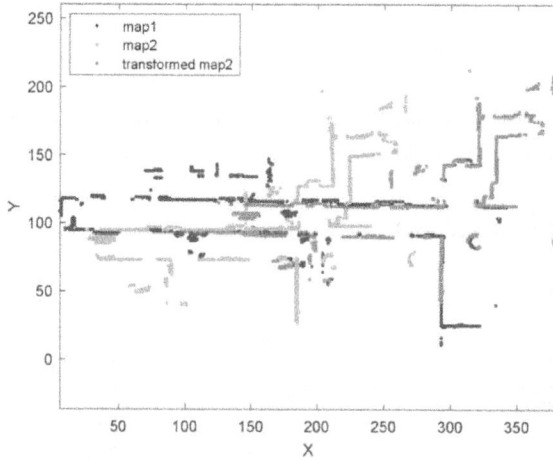

Figure 11.
The updated merged map by the ACO. The two individual maps were merged more accurately.

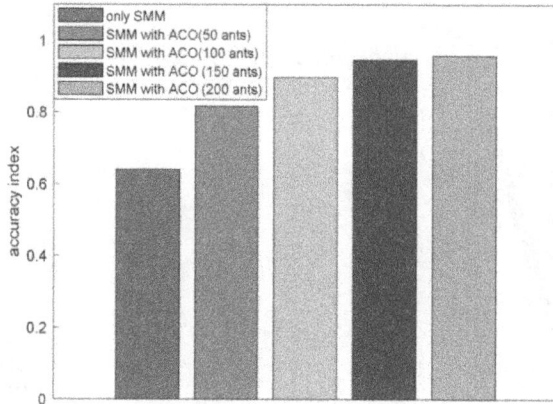

Figure 12.
The improved accuracy of the grid map merging with the ACO.

Although the accuracy index of the proposed method increases according to N_{ant}, the differences were not significant.

6. Conclusions

This chapter described how the ACO can be applied to the problem of grid map merging and analyzed how much the ACO improves the accuracy of grid map merging. The ACO needed to be modified to be applied to the merged map optimization. The search space for the ACO for grid map merging needs to be divided into two areas which contains the possible configurations of x and y translations respectively. The proposed method with the ACO was tested with the SMM which is a well-known indirect grid map matching algorithm. The ACO improved the accuracy of the SMM. The improved amounts increased slightly according to the number of ants in the ACO. Consequently, the modified ACO can be successfully applied to the problem of grid map merging and improve the accuracy of grid map merging.

Acknowledgements

This work was supported in part by the National Research Foundation of Korea (NRF) grant funded by the Korea government (MSIT) (No. 2019R1G1A1100597), and in part by the Grand Information Technology Research Center Program through the Institute of Information & Communications Technology and Planning & Evaluation (IITP) funded by the Ministry of Science and ICT (MSIT), Korea (IITP-2020-2020-0-01612).

Conflict of interest

There is no conflict of interest.

Author details

Heoncheol Lee
Department of IT Convergence Engineering, School of Electronic Engineering, Kumoh National Institute of Technology, Gumi, Gyeongbuk, South Korea

*Address all correspondence to: hclee@kumoh.ac.kr

IntechOpen

References

[1] Parker E, Bekey G, and Barhen J. Current state of the art in distributed autonomous mobile robots. Distributed Autonomous Robotic Systems. 2000;4;3–12. DOI: 10.1007/978-4-431-67919-6_1

[2] Lee H-C, Lee S-H, Choi M H, and Lee B-H. Probabilistic map merging for multi-robot RBPF-SLAM with unknown initial poses. ROBOTICA. 2012;30; 205-220. DOI: 10.1017/S026357471100049X

[3] Wikipedia. Lidar [Internet]. 2021. Available from: https://en.wikipedia.org/wiki/Lidar

[4] Civera J, Grasa O G, Davison A J and Montiel J M M. 1-Point RANSAC for EKF filtering: application to real-time structure from motion and visual odometry. Journal of Field Robotics. 2010;27;609-631. DOI: 10.1002/rob.20345

[5] Montemerlo M, Thrun S, Koller D and Wegbreit B. FastSLAM: A factored solution to the simultaneous localization and mapping problem. In: Proceedings of the AAAI National Conference on Artificial Intelligence; 28 July–1 August; 2002; Edmonton. Alberta. Canada. pp. 593–598

[6] Newman P and Ho K. SLAM-loop closing with visually salient features. In: Proceedings of the IEEE International Conference on Robotics and Automation; 18-22 April 2005; Barcelona. Spain. pp. 635-642

[7] Lee H. Tomographic feature-based map merging for multi-robot systems. Electronics. 2020;9;107(1–18). DOI: 10.3390/electronics9010107

[8] Lee H. Selective spectral correlation for efficient map merging in multi-robot systems. Electronics Letters, 2021 (Published online with early view). DOI: 10.1049/ell2.12139

[9] Rubinstein R Y, Ridder A and Vaisman R. Fast sequential Monte Carlo methods for counting and optimization; John Wiley & Sons: Hoboken, NJ, USA, 2013. DOI: 10.1002/9781118612323

[10] Kennedy J and Eberhart R. Particle swarm optimization. In: Proceedings of the IEEE International Conference on Neural Networks; 27 November–1 December; 1995; Perth. Australia. pp. 1942–1948

[11] Dorigo M and Gambardella L M. Learning approach to the traveling salesman problem. IEEE Transactions on Evolutionary Computation. 1997;1; 53-66. DOI: 10.1109/4235.585892

[12] Wikipedia. Ant colony optimization algorithms [Internet]. 2021. Available from: https://en.wikipedia.org/wiki/Ant_colony_optimization_algorithms

[13] Carpin. S. Fast and accurate map merging for multi-robot systems. Autonomous Robots. 2008;25;305–316. DOI: 10.1007/s10514-008-9097-4

[14] Mirjalili S. Ant Colony Optimization (ACO) [Internet]. 2021. Available from: https://www.mathworks.com/matlabcentral/fileexchange/69028-ant-colony-optimiztion-aco ; MATLAB Central File Exchange

Chapter 4

Ant Algorithms for Routing in Wireless Multi-Hop Networks

Martina Umlauft and Wilfried Elmenreich

Abstract

Wireless Multi-Hop Networks (such as Mobile Ad hoc Networks, Wireless Sensor Networks, and Wireless Mesh Networks) promise improved flexibility, reliability, and performance compared to conventional Wireless Local Area Networks (WLAN) or sensor installations. They can be deployed quickly to provide network connectivity in areas without existing backbone/back-haul infrastructure, such as disaster areas, impassable terrain, or underserved communities. Due to their distributed nature, routing algorithms for these types of networks have to be self-organized. Ant routing is a bio-inspired self-organized method for routing, which is a promising approach for routing in such Wireless Multi-Hop Networks. This chapter provides an introduction to Wireless Multi-Hop Networks, their specific challenges, and an overview of the ant algorithms available for routing in such networks.

Keywords: ant algorithms, wireless networks, ad hoc networks, mesh networks, wireless sensor networks, multi-hop

1. Introduction

Wireless Mesh Networks (*WMNs*) and Mobile Ad-Hoc Networks (*MANETs*) are applied in situations where there is no predefined network structure consisting of routers and base station or where the network is dynamic due to a growing number of nodes or mobile nodes moving into areas that have not been previously covered by a base station. Examples for such networks are Wireless Sensor Networks (*WSNs*) [1], vehicle ad-hoc networks [2], Wireless Senthe OLPC mesh network for children's computer in developing countries [3], and open grassroots initiatives to support free computer networks such as the Freifunk initiative in Germany [4] or the Funkfeuer initiative in Austria [5].

Routing of messages is a major challenge in such networks due to the dynamic or not *a priori* known network structure. Besides the problem of finding an optimum (or acceptable route) for a message, there is also a mutual influence of a used route on other routes. This calls for a self-organizing approach [6] of choosing routes that are near-optimal on a global level with a decision based on local information.

Artificial ant algorithms give a promising approach for such algorithms. Artificial ant algorithms are bio-inspired algorithms based on real ants' foraging behavior using a local gradient-following search strategy with pheromone trails. There are different ways how an ant-inspired algorithm can be implemented in Wireless Multi-Hop Networks, for example, by representing ants via network packets and the pheromone by values assigned to the network nodes. Besides the mapping of biological properties into a computer network, Algorithms differ in how route discovery and maintenance are implemented. This chapter investigates different ant algorithms and discusses their applicability to routing in wireless multi-hop networks.

The following section gives an introduction to ant-inspired emergence and self-organization. In Section 2, three forms of Wireless Multi-Hop Networks are introduced (*MANET*, *WSN*, or *WMN*). Section 3 describes first the seminal ant routing algorithm developed for routing in such networks, *AntHocNet*, followed by an overview of algorithms which build upon this algorithm. Section 4 provides a summary and concluding remarks.

2. Wireless Multi-Hop Networks

Recent advances in wireless communications have enabled the development of Wireless Multi-Hop Networks. Often also called Wireless Ad-Hoc Networks we will in the following use the term *Wireless Multi-Hop Network* to avoid confusion with (Mobile) Ad-Hoc Networks described in the next section. A Wireless Multi-Hop Network is a network where nodes communicate via several hops of wireless transmissions over equivalent nodes instead of a central base station. In this chapter we distinguish between (Mobile) Ad-Hoc Networks, Sensor Networks and Wireless Mesh Networks, which are described in the following.

2.1 (Mobile) Ad-Hoc Networks

An *Ad-Hoc Network* is defined as a network where all nodes communicate with one another on an ad-hoc basis without a central base station [7]. While sometimes the term is used in literature to denote a Wireless Multi-Hop Network, we use it here to denote a wireless network that does not differentiate between client nodes and dedicated routing nodes. The typical node is a somewhat powerful device such as a (ruggedized) laptop, smartphone, first-responder communication device, or a device that is integrated in a vehicle, all of which are possibly mobile (see **Figure 1**). In this case, the network is called a *Mobile Ad-Hoc NETwork* (*MANET*). These networks were originally developed for military use to enable troop communications in areas where no communications infrastructure was previously deployed. *MANETs* are also envisioned for emergency and disaster networking where the borders between *MANETs* and *WMNs* (see below) are flowing [8].

Several "classical" routing algorithms have been developed for *MANETs*, the most prominent being:

DSR: Dynamic Source Routing Protocol was developed in 1994 by David B. Johnson [9]. It is a *reactive* protocol which means that the protocol only builds routes on-demand. This is advantageous in highly volatile networks where it makes no sense to invest routing overhead to build and maintain routes that might go stale before they are used. On the other hand, traffic incurs a route-setup delay because the route is not built in advance. DSR is being standardized by the IETF [10],

DSDV: Destination-Sequenced Distance-Vector Protocol was developed by Charles Perkins and Pravin Bhagwat in 1994 [11]. It is a *pro-active* routing protocol, which means that routes are built in advance. This avoids the route-setup delay incurred by reactive protocols but, in highly volatile networks, a lot of routing overhead might be spent to set up and maintain routes that break before they can be used,

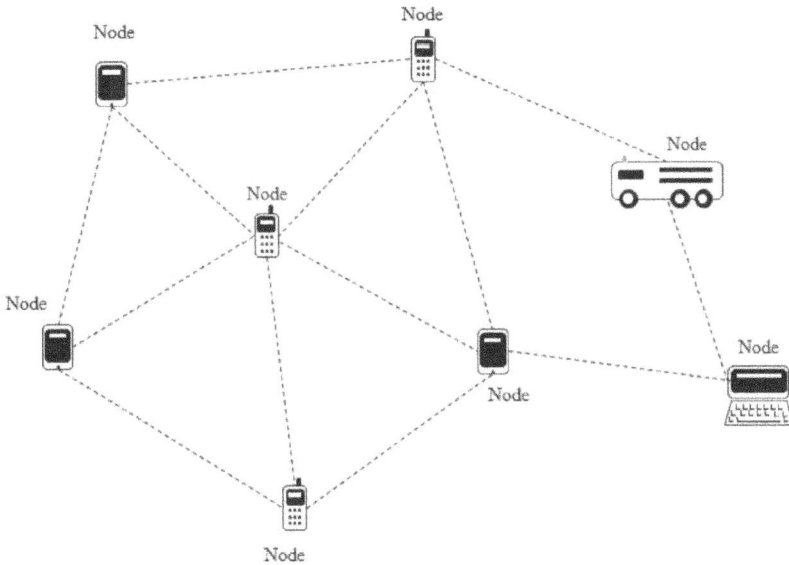

Figure 1.
MANET architecture. Nodes connect on an ad-hoc basis without dedicated routers or base stations.

AODV: Ad hoc On-Demand Distance Vector Routing [12] developed 1999 by
  Charles Perkins and Elizabeth Royer which is a reactive routing protocol, and
OLSR: Optimized Link State Routing Protocol [13] which was developed by
  Jacquet et al. in 2001. It is a proactive routing protocol and has been
  standardized by the IETF as experimental RFC 3626 [14].

2.2 Sensor Networks

A *Sensor Network* (or *Wireless Sensor Network*, *WSN*) is a wireless network that
consists of many sensor nodes which are spatially deployed to cooperatively monitor physical or environmental conditions (see **Figure 2**). *WSNs* were initially developed for the military for applications such as battlefield surveillance. However,
WSNs are now used in civilian application areas like environmental monitoring
[15–17]. In contrast to *MANETs* sensor nodes are typically tiny (so-called "motes")
and deployed densely in huge numbers, often on inaccessible terrain. Therefore,
sensor network protocols must be self-organizing.

One of the major problems in sensor networks is the limited power as each
sensor only has a small battery. While some sensors use technologies like solar cells
to refresh their batteries, routing protocols for sensor networks typically try to
optimize for power efficiency [18]. In *WSNs*, typically the sink initializes routing by
issuing a query for measurement data. Sensor nodes answer the query by sending
their data back to the sink. To save energy, data may be aggregated along the way
(*data-centric routing*). To facilitate this, the sensor field is often divided into clusters
or subnets. All nodes of a cluster first send their data to the respective cluster head,
which then processes and routes the data to the sink. If the sensors are equipped
with location finding devices like GPS (Global Positioning System), knowledge of
the position can be used to ease cluster formation or perform geo-routing. Well
known non-ant routing protocols for *WSNs* include:
  Gossiping is an early approach derived from flooding [19],
  SPIN: Sensor Protocols for Information via Negotiation is a family of protocols
    based on data-centric routing, developed by Heinzelman et al. in 1999 [20],

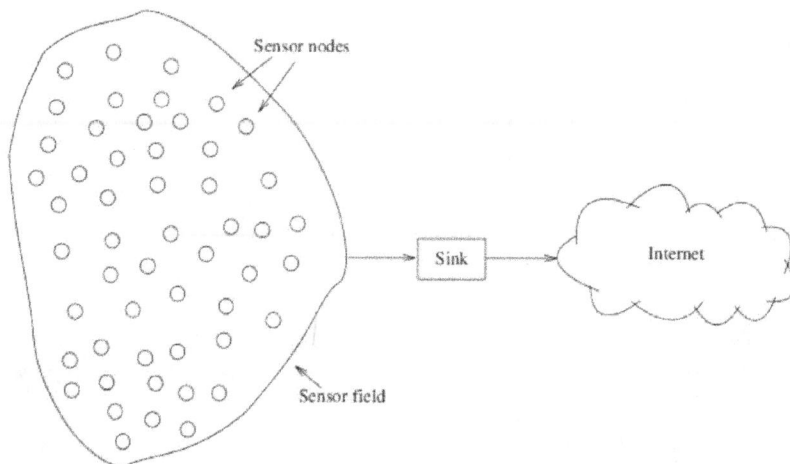

Figure 2.
WSN architecture. Sensor nodes are deployed in a sensor field and deliver their measurements through a sink to the Internet.

GPSR: Greedy Perimeter Stateless Routing in Wireless Networks, a geo-routing algorithm developed by Brad Karp in 2000 [21, 22],

LEACH: Low-Energy Adaptive Clustering Hierarchy developed by Heinzelman et al. in 2000 [23] which is a clustering-based protocol that minimizes energy dissipation,

SAR: Sequential Assignment Routing is an algorithm which selects paths based on energy resources, a QoS metric, and the packet's priority level. SAR was developed by Sohrabi et al. in 2000 [24] as part of a protocol suite for *WSNs*, and

Directed Diffusion, an approach using data-centric dissemination, reinforcement-based adaptation to the empirically best path, and in-network data aggregation and caching. Directed Diffusion was developed by Intanagonwiwat et al. in 2000 [25].

2.3 Wireless Mesh Networks

A *Wireless Mesh Network* (*WMN*) is a wireless networking architecture in which nodes are connected via a *wireless backbone* [26]. In contrast to *MANETs*, though, the wireless backbone in a *WMN* is typically fixed. Iow. a *WMN* consists of non-moving wireless mesh router nodes which constitute the wireless backbone and (potentially mobile) client nodes (see **Figure 3**). Router nodes can be mesh routers only (so-called Mesh Points, MP [27]) or act as combined *WLAN* Mesh Access Points (MAPs)/routers.

WMNs can be used as access networks to the Internet, where one or several Mesh Portal Points (MPP) connect the mesh to the Internet. They can also be used in disaster areas, for emergency response teams, and for the military. The boundary towards *MANETs* is somewhat fluid in these cases [8]. Consider, for example, a rescue operation where wireless routers are installed on top of firetrucks. When the firetrucks arrive at the scene, they stop and provide the wireless backbone for the firefighters' communication devices. This scenario can be seen as a *WMN* as well as as a *MANET*.

Even with a stationary wireless backbone, the characteristics of wireless channels and the interaction of the MAC layer with the higher layers in the network

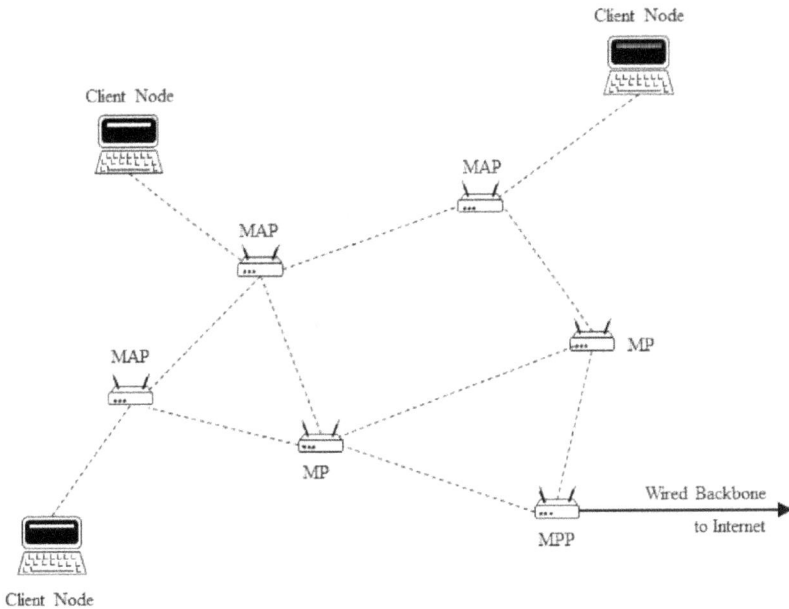

Figure 3.
WMN architecture. Client nodes connect via a wireless backbone. MP, Mesh Point; MAP, Mesh Access Point; MPP, Mesh Portal Point.

stack make routing in *WMNs* a hard problem. Wireless links vary over time and problems like the hidden node problem or the exposed node problem influence routing algorithm's performance. Therefore, a "wireless-aware" routing algorithm is necessary.

WMNs have slightly different constraints and pose different problems than *MANETs* and sensor networks. For example, the power constraint which is very prominent in sensor networks typically does not exist in *WMNs* [8]. When *WMNs* are used as access networks to the Internet "normal internet traffic" has to be assumed. This means application traffic like streaming, web browsing, VoIP (Voice over IP) and video conference traffic, or email, which use the standard TCP (or UDP) protocol stack. Routing algorithms for *WMNs* have mostly been adapted from the (Mobile) Ad-Hoc Networking Community.

3. Ant Colony Optimization and Ant-Routing Algorithms

Ant algorithms are inspired by the natural foraging behavior of certain species of ants. Based on the famous double bridge experiment, reported on by Goss, Aron, Denebourg and Pasteels in 1989 [28], ant-inspired optimization was then codified into an *Ant Colony Optimization* metaheuristic [29] which was originally implemented in algorithms such as *Ant System* [30] and *Ant Colony System* [31].

In general, algorithms using the *Ant Colony Optimization* metaheuristic work as follows: an optimization problem is transformed into a graph $G = (V, E)$, ants travel along the graph using pheromones (if present) to choose a path stochastically and after the ants have finished their travel, the pheromone values in the graph are updated according to the "goodness" of the solutions found by the ants. Many algorithm variants, also improve their results with a local search phase that is applied before updating the pheromone values. Besides other combinatorial optimization problems, these algorithms have been shown to be able to solve the traveling salesman problem. In *Ant System*, the first algorithm to implement the *Ant Colony Optimization* metaheuristic, the ants choose their path according to Eq. (1):

$$p_{ij}^k = \begin{cases} \dfrac{\tau_{ij}^\alpha \eta_{ij}^\beta}{\sum_{c_{il} \in N(s^p)} \tau_{il}^\alpha \eta_{il}^\beta} & \text{if} \quad c_{ij} \in N(s^p), \\ 0 & \text{otherwise} \end{cases} \tag{1}$$

where an ant k in a city i chooses the next city j with probability p_{ij}^k with s^p the partial solution constructed so far and $N(s^p)$ the set of possible edges leading only to cities not visited so far. The parameters α and β balance the importance of pheromone versus the local heuristic $\eta_{ij} = 1/d_{ij}$ with d_{ij} the distance between city i and city j.

Pheromones are updated using the update rule in Eq. (2):

$$\tau_{ij} \leftarrow (1 - \rho) \cdot \tau_{ij} + \sum_{k=1}^m \Delta \tau_{ij}^k. \tag{2}$$

with ρ the evaporation rate, m the number of ants and $\Delta \tau_{ij}^k$ proportional to the inverse of the lenght of the tour ant k took if that link was chosen (0 otherwise).

A variant aimed specifically at (wired) networks is *AntNet* [32]. These algorithms were not developed with Wireless Multi-Hop Networks in mind, though. As described above, Wireless Multi-Hop Networks have their own challenges in addition to the challenges of routing in a fixed network.

In the following, we will describe the seminal ant routing algorithm developed for Wireless Multi-Hop Networks, *AntHocNet* [33], and then give an overview of the typical features of other ant routing algorithms for these types of networks.

3.1 AntHocNet

AntHocNet [33], 2005, by Di Caro, Ducatelle, and Gambardella is the seminal ant algorithm developed for mobile ad-hoc networks. It addresses the special challenges that such wireless networks pose: bandwidth is typically less than in fixed networks, and links can change their quality or break. Therefore, *AntHocNet* is realized as a hybrid algorithm that combines features from pro-active and reactive routing protocols. In this way, it does not waste resources to set up paths before any packet is sent, which might not exist anymore by the time they are eventually needed.

Like all wireless routing algorithms, nodes running *AntHocNet* need to determine which other nodes are reachable by wireless transmission (iow. the one-hop neighborhood). *AntHocNet* nodes do this by broadcasting very short "hello" messages at regular intervals. Receiving nodes then set up these neighboring nodes in their respective routing tables, but without any routing information yet. These "hello" messages are also used to detect link failures.

When a new data packet is to be sent from a source node s to a destination node d, the algorithm enters its reactive path setup phase. There exist two possibilities: either there already exists routing information for a path between s and d (after the protocol has run for a while and packets have already been sent) or not. Depending on whether routing information already exists or not, *AntHocNet* sends so-called "forward ants" either by broadcasting them (if no routing information for the required route exists yet) or by unicasting them stochastically along one of the already known routes. For unicasting, the pheromone routing tables at the intermediate nodes are exploited and the next hop n towards the destination d is chosen stochastically with a probability P_{nd} according to Eq. (3):

$$P_{nd} = \frac{\left(T_{nd}^i\right)^{\beta}}{\sum_{j \in \mathcal{N}_d^i}\left(T_{jd}^i\right)^{\beta}}, \beta \geq 1. \tag{3}$$

with i the current node, n the next hop, T the respecitve pheromone value, $\mathcal{N}_d^i$ the set of possible neighbors and a coefficient β that controls how explorative the algorithm behavior is.

If there is no pheromone information available yet, the forward ant is broadcasted. To avoid flooding the network with too much traffic, these broadcast ants are restricted in several ways: 1) after a number of hops, they are killed, 2) when a node receives several ants stemming from the same broadcast (that took different paths to reach this node), it will only pass on those ants which came via sufficiently good paths (using the number of hops and travel time as metrics). The threshold for this can be set by a parameter α_1. In this way, several parallel paths can be explored while the worst paths are quickly excluded and overhead (which is always of special concern in wireless networks) is kept at a reasonable level. A second parameter α_2 is used to spread paths more widely among the network: broadcast ants which took a different first hop than previous ants stemming from the same broadcast, this less restrictive parameter α_2 is applied instead of α_1.

Ants memorize the path they travel and when a forward ant has reached the destination node, a so-called backward ant is created which travels back the path $P = s \rightarrow n_1 \rightarrow n_2 \rightarrow \dots \rightarrow d$ it came. This backward ant then updates all the pheromone information along the path according to Eq. (4):

$$T_{nd}^i = \gamma T_{nd}^i + (1-\gamma)\tau_d^i, \quad \gamma \in [0,1]. \tag{4}$$

where τ_d^i is an expression of the "goodness" of the path, based on an estimate of the average time to send a packet over the path P calculated from measurements at each node's MAC layer.

After one or several path(s) has been found and while a data session is running, *AntHocNet* forwards the data packets stochastically along all the available paths using the same Eq. (3) as the forward ants but with a higher value of β. This means that data packets have a more exploitative behavior than ant packets which explore more.

During a data session, *AntHocNet* enters its pro-active phase and sends forward ants in addition to the data packets. These again use Eq. (3) but have a small probability of being broadcast instead. The ants that follow the existing path via Eq. (3) update the current quality of the existing path while those ants that are broadcast can potentially find new, better paths which will then be immediately used as potential paths to route data. Due to the way paths are determined and updated during the pro-active phase and due to the stochastic nature of the data routing, data packets are sent in an automatically load-balanced way through the network which expecially helps with wireless transmission as two parallel paths use the same transmission medium and therefore can potentially greatly influence each other.

AntHocNet also addresses link failures, which occur much more frequently in the wireless domain. As mentioned before, link failure can be detected via "hello" messages – if there has not been an "hello" message for a certain amount of time (several times the regular sending interval), a link to a node will be considered broken. A link is also considered broken if a unicast message to a node fails. The algorithm then enters its local path repair mode where it broadcasts so-called "path repair ants" that work just like the forward ants in reactive mode, except that they are more limited in their maximum number of allowed broadcasts. If a path can be repaired within a certain amount of time, the data packets (which will have been buffered in the meantime) will be sent to the destination node. If the path can not be re-established within a reasonable time limit, the data is discarded and link failure notifications are broadcasted to the surrounding nodes.

3.2 ARA

ARA (Ant-Colony-based Routing Algorithm) was proposed for *MANETs* by Güne s, Sorges and Bouazizi in 2002 [34]. *ARA* is based on a version of *Ant Colony Optimization* which the authors call "Simple ACO". Its main goal is to reduce the overhead of routing as compared to classical routing algorithms. The algorithm consists of three phases: route discovery, route maintenance, and route failure handling. It uses routing tables at each node n_i which consist of records of the form $\langle n_d, n_n, \varphi_{i,n} \rangle$ where n_d is the destination node, n_n the next hop node and $\varphi_{i,n}$ the pheromone value for this link and destination. Ants carry a sequence number and the source address they originated from.

The transition rule looks very much like that of *AntHocNet* (cf. also Eq. (3)) with a coefficient of $\beta_1 = 1$:

$$p_{i,n} = \begin{cases} \dfrac{\varphi_{i,n}}{\sum_{j \in N_i} \varphi_{i,j}} & n \in N_i \\ 0 & n \notin N_i \end{cases} \tag{5}$$

where $p_{i,n}$ is the transition probability of going from the current node n_i to node n_n and N_i is the set of one-hop neighbors of n_i.

In contrast to *AntHocNet*, though, ants are used differently as follows:

- During route discovery, forward ants do not follow the transition rule but are broadcasted instead. Duplicate forward ants can be detected by their sequence number and source address and are not forwarded.

- On its way through the network, the forward ant immediately updates the pheromone tables at the nodes – for the way back to the source. The forward ant is interpreted similarly to backward ants in *AntHocNet*. When a forward ant arrives at a node, the routing table entry where n_d equals the source address in the ant and n_n equals the last hop the ant took is updated. Backward ants are used analogously and establish the path to the destination node as usual.

The pheromone update rule in *ARA* is quite simple; an ant changes the pheromone value moving from node n_i to n_n by a constant amount:

$$\varphi_{i,n} := \varphi_{i,n} + \Delta\varphi. \tag{6}$$

Güneş et al. suggest that the number of hops an ant has traveled to the current node could also be included in the calculation of the new amount of pheromone. Pheromone is evaporated in regular intervals according to the evaporation rule shown in Eq. (7). The authors suggest that the link quality measurements should be incorporated into the evaporation rule rather than into the pheromone update rule as usual. This has the advantage that nodes can update local changes of link quality much more quickly. On the other hand, the disadvantage of this method is that the quality reflected by the amount of pheromone reflects local link quality only instead of end-to-end path quality.

$$\varphi_{i,n} := (1 - q) \cdot \varphi_{i,n}, \qquad q \in (0, 1] \tag{7}$$

Route maintenance is done by observing the traffic flowing through the network. Traffic does not have to be encapsulated in ant packets; rather, nodes autonomously update the pheromone tables according to the pheromone update rule already shown in Eq. (6). For each packet of traffic observed by the node, the pheromone value is increased by the constant amount $\Delta\varphi$. This has the advantage that route reinforcement happens "automatically" without the need for extra ant packets.

To prevent the creation of loops, *ARA* implements a simple loop avoidance mechanism. When a node recognizes the duplicate reception of a data packet (identifiable by sequence number and address), it sets an error flag and sends the packet back to the previous node which removes the link from its routing table.

Route failures are recognized by missing acknowledgements. When a link fails, a node first checks whether it has another route to the required destination in its routing table. If this is the case, it sends the packet via this alternative link. If not, the node informs its neighbors anticipating that they can relay the packet. Failing this, the mechanism tracks back until it arrives back at the source node. In that case, a new route discovery phase has to be initiated by the source node.

3.3 ARAMA

Ant Routing Algorithm for Mobile Ad-hoc networks (*ARAMA*) was published by Hussein, Saadawi and Lee in 2005 [35]. It is targeted at *MANETs* and *WSNs* and focuses on fair resource usage – esp. node energy – across the network. To achieve this, the forward ants carry not only source and destination address and intermediate node IDs but also quality information about the path. To prevent ants from growing too big, the path information is calculated as a normalized local index and computed into a cumulative path index as shown in Eqs. (8) and (9) below. The ant only carries the path index. This novel path index is the main contribution of the paper.

Let $p_{i,m}$ node i's normalized optimization parameter m with $0 < p_{i,m} < 1$. This can be the number of hops, battery power, delay, bandwidth, etc. Then the local normalized index I_i for node i is

$$I_i = \sum_m a_m p_{i,m} \tag{8}$$

where a_m is the weight of this parameter with $\sum_m a_m = 1$. This leads to $0 \le I_i \le 1$. As the forward ant passes a node it updates the path information it carries by calculating the path index I_{path} as follows:

$$I_{path} = \prod_i I_i. \tag{9}$$

Since $0 \le I_i \le 1$ also $0 \le I_{path} \le 1$. A bottleneck link on the path correctly influences the overall path index as the value of I_{path} is smaller than the smallest I_i along the path. When the forward ant reaches the destination, the path grade ρ is computed as

$$\rho = f\left(I_{path}, I_{path,best}\right) \tag{10}$$

where $I_{path,best}$ is the best I_{path} received in the last W number of ants (W a suitable window size).

With d the destination node, i the current node, and n the next hop node the **transition rule** used by *ARAMA* is given as

$$p_{d,i,n} = \begin{cases} \dfrac{fun\left(\tau_{d,i,n}, \eta_{i,n}\right)}{\sum_{j \in N_i} fun\left(\tau_{d,i,j}, \eta_{i,j}\right)}, & n \in N_i \\ 0 & n \notin N_i \end{cases} \tag{11}$$

where $\tau_{d,i,n}$ is the pheromone value for going from current node i to destination d via next neighbor n and $\eta_{i,n}$ is the local heuristic value of the link (i,n). Function $fun\left(\tau_{d,i,n}, \eta_{i,n}\right)$ is chosen to give a high function value when $\tau_{d,i,n}$ and $\eta_{i,n}$ are high; eg. as in the transition rule of *AntHocNet*.

When the backward ant traverses the network back to the source, the pheromones are updated with the **pheromone update rule**:

$$\tau_{d,i,n} := \begin{cases} f_{evap}(\rho_d)\tau_{d,i,n}) + g_{enf}(\rho_d) & \text{if} \quad n \in \text{Path} \\ f_{evap}(\rho_d)\tau_{d,i,n}) & \text{if} \quad n \notin \text{Path} \end{cases} \tag{12}$$

with $f_{evap()}$ the evaporation function, g_{enf} the enforcement function, and ρ_d the path grade for this path calculated from the information in the forward ant as shown above.

The authors also propose two very interesting extensions to the algorithm:

Negative Backward Ants are sent if a forward ant dies due to running out of TTL (time-to-live) or loop detection. In this case, a negative backward ant is sent which deemphasizes the path by decreasing its pheromone levels.

Destination Trail Ants implement the *RARE* (Receiver Assisted Routing Enhancement) concept by the same group (Abdelmalek, Hussein, and Saadawi [36]). With this technique, destination nodes send so called "destination trail ants" into the network which randomly mark paths leading to the destination. When forward ants search for this destination there is a probability that they will hit a destination trail left behind by a destination trail ant. This helps to speed up connection setup time.

3.4 AMQR

Ant colony based Multi-path QoS-aware Routing (*AMQR*) was developed for *MANETs* by Liu and Feng in 2005 [37]. It is based on *ARA* (introduced in Section 3.2). In contrast to *ARA* it supports link-disjoint multi-path routing.

Like *ARA* it uses the transition rule from *AntHocNet* with a factor $\beta_1 = 1$ (cf. Eq. (3)).

$$p_{i,n} = \begin{cases} \dfrac{\varphi_{i,n}}{\sum_{j \in N_i} \varphi_{i,j}}, & \text{if} \quad n \in N_i \\ 0 & \text{if} \quad n \notin N_i \end{cases} \tag{13}$$

with $p_{i,n}$ the probability to go from node n_i to n_n and N_i the set of neighbors of n_i in dependence of the pheromone value $\varphi_{i,n}$.

As in *ARA*, forward ants mark the trail back to the source as they move while backward ants update the trail to the destination. The pheromone values are updated as follows:

$$\varphi_{i,n} := (1 - \alpha) \cdot \varphi_{i,n} + \Delta\varphi_{i,n} \tag{14}$$

with α the pheromone decay parameter and $\Delta\varphi_{i,n}$ calculated as

$$\Delta\varphi i, n = q^{-m} h^{-n} \tag{15}$$

where q is the delay time and h the hop count experienced by the ant so far. The parameters m and n are the weights that determine the relative importance of time delay and hop count.

Forward ants use a frame format of $\langle n_s, n_d, SeqN, HopC, [(PasN, ArrT), (.,.), ...] \rangle$. where n_s is the ID of the source node, n_d the ID of the destination node, and *SeqN* and *HopC* the sequence number and hop count of the ant respectively. The list $[(PasN, ArrT), ...]$ contains the IDs of the nodes *PasN* passed by the ant and the relevant arrival times *ArrT*. Backward ants use the same frame format as forward ants but without *SeqN* and *HopC*.

The routing table has the usual entries $\langle n_d, n_n, \varphi_{i,n} \rangle$ with n_d the destination, n_n the next hop and $\varphi_{i,n}$ the pheromone value for this link.

During route discovery, a source node first sends hello packets to determine its neighbors and then broadcasts a forward ant. Therefore, there is more than one copy of the forward ant in the system. When an intermediate node receives a forward ant more than once and the ant's hop count $HopC_{new} \leq HopC_{old} + \Delta hops$ another entry is made in the routing table to record this alternative path. Parameter $\Delta hops$ is the threshold for an acceptable additional path length to avoid overly long alternative paths. Backward ants always choose the best path back to the source.

Nodes exchange routing information by additional communication and build their own view of the topology, and only link-disjoint routes are used. The same concept is used in *PPRA* shown later (see Section 3.6). Load balancing and route failures are handled as in *ARA*.

To support QoS, nodes monitor their state and the delay recorded in the ants they receive. If the delay in an ant exceeds a certain limit, the pheromone for the respective link is set to 0 and the other pheromone values in the routing table are adjusted to eliminate this high-delay link. If the node itself is overloaded, it initiates a new backward ant to the source to change the route.

3.5 Scalable Ant-based Routing

Ohtaki et al. [38] focus on the scalability of ant-based routing for *MANETs*. Their algorithm is based on uniform ant routing [39] and borrows the TTL-limiting technique from HSLS (Hazy Sighted Link State) routing [40].

As in uniform ant routing, a probability routing table is kept at each node. For each entry (d, n) there exists a value p_{dn} which gives the probability of routing a packet destined for node d via neighboring node n. Nodes send periodic control messages (ants) of the form $\langle h_s, c, TTL \rangle$ which wander the network randomly. Here, h_s is the source address, c the cost of all links traversed so far, and *TTL* the remaining time-to-live. Whenever a node receives such a control message from a neighboring node l it updates its routing table as follows

$$p_{dn} = \begin{cases} \dfrac{p_{dn} + \Delta p}{1 + \Delta p}, & \text{if } n = l \\[3mm] \dfrac{p_{dn}}{1 + \Delta p} & \text{if } n \neq l \end{cases} \qquad (16)$$

with $n \in N_i$ the set of neighbors of the current node i and Δp given by

$$\Delta p = \frac{k}{f(c)}, (k > 0) \qquad (17)$$

where $f(c)$ is a function of the total cost c and k the so-called learning rate of the algorithm. It defines the weight of one ant and is generally less than 0.1.

As the number of nodes in the network increases, the number of ants goes up and becomes a burden on the network. To improve scalability, the Scalable Ant-based Routing algorithm borrows a technique from HSLS [40] to limit the *TTL* of the control packets. The *TTL* T_k of the k-th ant is calculated as

$$T_k = 2^{x_k + 1} \qquad (18)$$

where

$$x_k = \min\left(x_{max}, \max\left(x | k \equiv 0 (\bmod 2^x)\right)\right). \qquad (19)$$

The authors suggest that x_{max} should be set to half the number of nodes in the network.

Another improvement of this algorithm is the novel ant migration scheme. Instead of a purely random walk, ants try to move as far away from the source as possible. The idea is that ants should not "waste" their *TTL* in the neighborhood but rather try to cover the whole network. They can find the "direction away from the source" by following those links which have a low probability as a way to the source in the routing table. Iow. when an ant originated from source s was received from node m the probability q_j that node j will be chosen as next-hop node among all neighboring nodes except m is calculated as

$$q_j = \frac{\frac{1}{p_{sj}}}{\sum_{k=1}^{n} \frac{1}{p_{sk}} - \frac{1}{p_{sm}}} \qquad (20)$$

to find the next link in direction away from the source. In this way, they use their TTL most efficiently to reach nodes as far away from the source as possible and get good coverage of the network.

3.6 PPRA

PPRA Prioritized Pheromone Aided Routing Algorithm was published by Jeon and Kesidis in 2005 [41]. It is a multipath routing algorithm that considers both energy and latency and supports dual-priority traffic. It is aimed at sensor networks (*WSNs*) and *MANETs* with battery constraints and based on *ARA* [34] (see also Section 3.2). Multipath routes are used to guard against route failures; in case of a route breaking, the already set-up backup route can be used without waiting for another route discovery phase. The multiple paths are also used for load balancing. As the primary path's pheromones degrade, traffic switches to the alternate routes.

The routing tables at each node n_i consist of entries of the form $\langle n_d, n_n, \delta^i(d,n), e^i(d,n)\rangle$ or $\langle n_d, n_n, \partial^i(d,n), e^i(d,n)\rangle$ where n_d is the destination node, n_n the next hop node, $\delta^i(d,n)$ the TTL-pheromone or $\partial^i(d,n)$ the Delay-pheromone respectively, and $e^i(d,n)$ the Energy-pheromone described later.

During **route discovery** the forward ants are broadcasted. Like in source routing, they carry the source address and record all node addresses along the way. Duplicate forward ants are *not* discarded as in single path algorithms. Instead, they are used to set up alternative routes. Note that out of the paths found by the duplicate ants only those which are link-disjoint are kept. Once a forward ant reaches the destination, a backward ant is created, which takes the path found by the forward ant back to the source. For the measurement of energy and delay, periodic control packets are sent in addition to route discovery ants.

Pheromone types: There are three kinds of pheromones in *PPRA*: TTL-pheromone (δ), Delay-pheromone (∂), and Energy-pheromone (e). The algorithm has two variants. Variant 1 uses TTL-pheromone and Energy-pheromone, while Variant 2 uses Delay-pheromone and Energy-pheromone.

TTL-pheromone is used to express the distance in hops (iow. the time-to-live a packet traveling this path will use) and is calculated as

$$\delta^i(d,n) := \delta^i(d,n) + \beta_1 \cdot TTL(d,n) \tag{21}$$

with β_1 a scaling constant and $TTL(d,n)$ the number of hops between node d and node n. Evaporation for TTL-pheromone is calculated as

$$\delta^i(d,n) := \delta^i(d,n) \cdot \beta_2 \quad \text{with} \quad 0 < \beta_2 < 1. \tag{22}$$

For highly volatile networks a higher value of β_2 can be used to decay stale routes faster.

Energy-pheromone represents the battery status of the nodes in the path. Similar to Eq. (21) it is calculated as

$$e^i(d,n) := e^i(d,n) + \alpha_1 \cdot E_{min}(d,n) \tag{23}$$

with α_1 a scaling constant. E_{min} represents the energy bottleneck on the path, i.e. the lowest battery level encountered in a node along the path. Evaporation for Energy-pheromone is calculated as

$$e^i(d,n) := e^i(d,n) \cdot \alpha_2 \quad \text{with} \quad 0 < \alpha_2 < 1. \tag{24}$$

Delay-pheromone marks the cumulative queuing delay experienced along a path. It is calculated analogously to TTL-pheromone as

$$\partial^i(d,n) := \partial^i(d,n) + \gamma_1 \cdot D(d,n) \tag{25}$$

with γ_1 a scaling constant and $D(d,n)$ the cumulative queuing delay between node d and node n. Evaporation for Delay-pheromone is calculated as

$$\partial^i(d,n) := \partial^i(d,n) \cdot \gamma_2 \quad \text{with} \quad \gamma_2 > 1. \tag{26}$$

The algorithm distinguishes between latency-critical and non-critical traffic. For latency-critical traffic, it always uses both pheromone levels (Energy- and TTL-pheromone or Energy- and Delay-pheromone respectively) to determine the route, for non-critical traffic only Energy-pheromone is considered.

The **transition rule for non-critical traffic** is given in Eq. (27).

$$p_e^i(d,n) = \frac{e^i(d,n)}{\sum_{j \in N_i} e^i(d,j)} \tag{27}$$

where N_i the set of neighboring nodes of i.

The **transition rule for latency-critical traffic** is calculated by combining Energy-pheromone and TTL-pheromone (algorithm variant 1) or Energy-pheromone and Delay-pheromone (algorithm variant 2) as shown in Eqs. (28)–(31) respectively.

$$\text{Variant 1}: \quad p_{lat}^i(d,n) = \frac{\theta \cdot p_e^i(d,n) + p_\delta^i(d,n)}{\sum_{j \in N_i} [\theta \cdot p_e^i(d,j) + p_\delta^i(d,j)]} \tag{28}$$

where

$$p_\delta^x(w,z) = \frac{\delta^x(w,z)}{\sum_{j \in N_x} \delta^x(w,j)} \tag{29}$$

and N_x the set of neighboring nodes of x.

$$\text{Variant 2}: \quad p_{lat}^i(d,n) = \frac{\theta \cdot p_e^i(d,n) + p_\delta^i(d,n)}{\sum_{j \in N_i} [\theta \cdot p_e^i(d,j) + p_\delta^i(d,j)]} \tag{30}$$

where

$$p_\partial^i(d,n) = \frac{1/\partial^i(d,n)}{\sum_{j \in N_i} 1/\partial^i(d,j)} \tag{31}$$

and N_i the set of neighboring nodes of i.

3.7 EEABR

Energy-Efficient Ant-Based Routing (*EEABR*) is a routing algorithm based on the *Ant Colony Optimization* metaheuristic. It was developed for *WSNs* by Camilo et al. in 2006 [42]. The major goal of this algorithm is to increase energy efficiency. The authors propose three algorithms, basic, improved, and energy-efficient ant routing.

$$p_k(r,s) = \begin{cases} \dfrac{[\tau(r,s)]^\alpha [E(s)]^\beta}{\sum_{u \notin M_k} [\tau(r,u)]^\alpha [E(u)]^\beta}, & \text{if } s \notin M_k \\ \\ 0 & \text{if } \textit{otherwise} \end{cases} \tag{32}$$

where an ant k chooses with probability $p_k(r,s)$ to move from node r to node s, $\tau(r,s)$ the amount of pheromone for link (r,s), and $E(s)$ being the factor η in the *Ant Colony Optimization* metaheuristic. In this case, $E(s)$ is calculated from the initial energy level of the nodes C and e_s the actual energy level of the node by

$$E(s) = \frac{1}{C - e_s}. \tag{33}$$

The backward ant of forward ant k drops pheromone according to the pheromone update rule given in Eq. (34).

$$\tau_k(r,s) := (1 - \rho) \cdot \tau_k(r,s) + \Delta\tau_k. \tag{34}$$

Here, $(1 - \rho)$ represents the evaporation and $\Delta\tau_k$ is calculated from the total number of nodes N and the distance Fd_k traveled by forward ant k as

$$\Delta\tau_k = \frac{1}{N - Fd_k}. \tag{35}$$

The first improvement of this basic algorithm uses a refined function for calculating τ_k where the ant carries an energy vector. From this, the average energy level of the path is calculated when the backward ant is created. While this makes it possible to better monitor the energy level on the path it can lead to quite big forward ants. Since in *WSNs* communication costs much more energy than local calculation, the authors propose to save ant size by storing only the average $(E\,avg_k)$ and minimum energy $(E\,\min_k)$ found on the path so far in the forward ant. The pheromone update rule in the final *EEABR* algorithm is then given as:

$$\tau_k(r,s) := (1 - \rho) \cdot \tau_k(r,s) + \left[\frac{\Delta\tau_k}{\varphi Bd_k}\right] \tag{36}$$

where φ is a coefficient, Bd_k the traveled distance of the backward ant in hops and

$$\Delta\tau_k = \frac{1}{C - \left[\frac{E\,\min_k - Fd_k}{E\,avg_k - Fd_k}\right]}. \tag{37}$$

Through factors φ and Bd_k the backward ant loses part of its pheromone strength while it travels back to the source – thereby giving shorter paths an advantage in the routing table.

The authors also reduce the memory M_k of already visited nodes in the forward ant to just the last two nodes visited. This means no full path information is stored in the ant anymore, further reducing its size to achieve a so-called "light-weight" ant. The tasks of loop detection and remembering the path back to the source now fall to the nodes themselves. Nodes keep track of the forward ants using a structure $\langle n_p, n_s, ant_{\text{ID}}, t \rangle$ where n_p is the previous node, n_s the next (forward) node, ant_{ID} the ID of the ant and t a timeout value.

When a forward ant is received, a node checks the table whether this ant has been received before. If yes, the ant is discarded as a loop was detected. If no, the ant is forwarded according to the transition rule. When the backward ant returns to the node, it looks up the way back to the source in this same table. The timeout timer t controls how long the node keeps the entry in the table. This also determines the maximum time a backward ant may take to come back via this node.

3.8 DDCHA

The Distributed, Data-Centric, Hierarchical Ant algorithm (*DDCHA*) is a combination of a data-centric protocol with ants developed for *WSNs*. To aggregate data, the sensor field is divided into subnets where the biggest distance between nodes is still within communication range. Nodes are location-aware and join a subnet based on their location relative to the sink. In each subnet, a core head and a gateway are chosen with the Distributed Energy-Core Generating Algorithm (DECGA) described in the same paper as follows:

1. Initially, the sink node is the core head, and all sensor nodes are member nodes.

2. Every node exchanges information about its function (core head, gateway, member) and energy level with its neighbors periodically.

3. In every period, every node p computes its new state

   - If there is no core head in a subnet, then the node with the largest surplus energy becomes the core head.

   - If p is neither a core head nor gateway but neighbor with at least one node of a different subnet then p becomes a gateway.

The authors prove that this generates an energy-core Ψ in the network graph.

Routing is done with the *DDCHA* ant algorithm on top of this network structure as follows: initially, all pheromone values are 0. Every core head can be seen as an ant nest (source) which sends forward ants towards the sink of the *WSN*. Forward and backward ants both mark their path with pheromones immediately. Unlike *ARA* (and closer to ant behavior in nature), the ants do not mark the path back to the source/destination but simply drop a fixed amount $+\Delta$ of pheromone on the forward path. Ants also do not follow a probabilistic routing table but choose the path with the highest amount of pheromone. If an ant can not move on anymore (i.e., it got caught in a loop) it backtracks its path, decreasing the amount of pheromone by $-\Delta$ on the way until it finds a new path. Loop detection is achieved by keeping a forbidden-list of already visited nodes in the ant. Each member node sends its data to the core head first which aggregates the data. The core head then sends the data onto the sink of the *WSN*.

3.9 Approaches using colored pheromones

Several algorithms are known which make use of "colored pheromones". This means that trails can be distinguished not only by the amount of pheromone dropped but also by the "color" of the pheromone. In the following, we introduce three approaches for *MANETs*, *WSNs*, and *WMNs* respectively.

3.9.1 MACO

The original *Ant Colony Optimization* metaheuristic has some drawbacks in terms of stagnation and adaptiveness. Stagnation occurs when a network reaches convergence, and an optimal path is found and chosen by all ants. This, in turn, reinforces this path so much that the probability of selecting other paths becomes very low, which can lead to congestion on the "optimal" path. Adaptiveness describes the ability of an algorithm to react to changes in the network. *MACO* (Multiple Ant Colony Optimization) was developed to mitigate these problems in *MANETs* by Sim and Sun in 2002 [43].

MACO uses several ant colonies in parallel, which each use their own color of pheromone. The colonies are entirely separated and cannot sense pheromone other than that of their own color.

Similar to nature, the forward ants immediately drop pheromone on the paths they take. I.e., a "red" forward ant A^r will drop $+\tau$, on the forward direction of a link it chooses. The backward ant inherits the color from the forward ant and chooses the path back to the source with the highest amount of pheromone in its respective color. Backward ants also drop additional pheromone on the link on their way back.

Depending on the number of ant colonies used in parallel, several paths from the source to the sink can be found. These paths can then be used as alternative routes for load-balancing. Consider the following example with two ant colonies, "red" and "blue". Lets assume that there are three paths through the network, one long route R_1 and two similarly short (=good) ones R_2, R_3 with $R_1 > R_2 \approx R_3$. With just one ant colony, the true minimum route would be chosen as the optimal path. With two ant colonies, there is a possibility that one colony will find R_2 as the shortest path and the other colony R_3. In this case, two alternative paths of similar quality were found. Therefore, *MACO* increases the probability of exploring alternatives.

3.9.2 Division of labour in SANETS

Wireless *SANETs* (Sensor/Actuator Networks) are a form of *WSN* which also contains actuators (eg. robots). The same energy constraints as in *WSNs* apply.

Labella and Dressler develop in [44] an ant-based algorithm for division of labour and the routing of the respective traffic in *SANETs*. In their model, nodes can perform different tasks (measurement of temperature, recording of sound, recording of video, and movement). The goal is to distribute the tasks evenly in the network to get good measurement coverage and not overload single paths with high-load communications (video and audio transmissions).

Nodes choose tasks by employing the *AntHocNet* transition rule. The probability for a node to choose task i from the task list T_{agent} is

$$P(i) = \frac{\tau_i^{\beta_{\text{task}}}}{\sum_{k \in T_{\text{agent}}} \tau_k^{\beta_{\text{task}}}}. \tag{38}$$

Pheromones are updated depending whether the task could be completed successfully or not by

$$\tau_i := \min \tau_{max}, \tau_i + \Delta\tau \tag{39}$$

if successful and

$$\tau_i := \max \tau_{min}, \tau_i - \Delta\tau \tag{40}$$

if not.

Since tasks are inherently linked with the traffic they generate (simple temperature values, sound, video, or command traffic for movements) tasks imply traffic classes. To deal with these different classes, the authors employ colored pheromones.

Each node i keeps separate routing tables $_c\mathbf{R}^i$ for each color c. Each entry $_c\mathbf{R}^i_{nd}$ for going from node i via node n to destination d is of the form $\langle t, h, e, s, m, v \rangle$ where t is an estimation of the transmission time, h the number of hops in the route, e the energy required for transmission, s the minimal signal-to-noise ratio on the path, and m and v are flags that indicate whether a node is mobile and still valid for routing.

The transition rule is taken from *AntHocNet* and evaluated for each color:

$$\mathcal{P}^i_{nd} = \frac{r\left(_c\mathbf{R}^i_{nd}\right)^\beta}{\sum_{j \in N^i_d} r\left(_c\mathbf{R}^i_{nd}\right)^\beta} \tag{41}$$

where N^i_d is the set of neighbors for which a path to d is known. A different value for β is used during route discovery and data routing. $r(\cdot)$ is a function $r : \mathbf{R} \to \mathbb{R}^+$ which evaluates the link statistics.

The algorithm also employs an elaborate probabilistic scheme to filter packets and deliberately drop them to avoid congestion (eg. if a measurement value did not change).

3.10 Other Ant-based Algorithms and Techniques

Other ant-based algorithms for Wireless Multi-Hop Networks include the following:

AARAI: Ant Colony Algorithm with Adaptive Improvement was developed by Zeng and He in 2005 [45]. It is targeted at *MANETs*, based on *Ant Colony Optimization* and implements multipath routing.

IAQR: Improved Ant colony QoS Routing, developed by Liu et al. in 2007 [46] is based on *Ant Colony System* [31] and has the goal to improve QoS routing in *MANETs*. The transition rule is taken from *AntHocNet*. It measures each link delay, bandwidth, jitter, and cost and for each node queueing delay, packet loss, jitter, and cost. It uses a global update rule for QoS optimization which changes the decay factor of the pheromone to account for link quality.

Ant-AODV: Ant-Ad Hoc On-Demand Distance Vector Routing by Marwaha et al. [47] combines AODV [12] with ants. It is targeted at *MANETs*. Ants are used during the route discovery phase to reduce route discovery latency.

Zhang, Kuhn, and Fromherz [48] introduce three new ant routing algorithms for *WSNs*: Sensor-driven and Cost-aware ant routing (SC), Flooded Forward ant routing (FF), and Flooded Piggybacked ant routing (FP). In SC, they introduce the concept of "sensing ants" which can "smell" other nodes from afar. This is facilitated by exchange of cost estimates from neighboring nodes via HELLO-messages. In FF, forward ants are flooded using the broadcast channel of the *WSN* and in FP forward ants and data ants are combined to save communications overhead.

Other techniques use ants not directly to perform routing but to support the actual routing algorithm. Examples of these are described in the following.

GPSAL: GPS/Ant-Like Routing Algorithm by Câmara and Loureiro [49, 50] is a location based routing algorithm designed for *MANETs* and *WMNs*. Its goal is to use location information to reduce the number of routing messages and speed up route recovery. The assumption is that nodes are equipped with location finding devices like GPS (Global Positioning System). Nodes n_i keep routing tables which contain $\langle n_d, loc_{curr}(d), loc_{prev}(d), TTL\left(loc_{prev}(d)\right), type(d)\rangle$ where n_d is the destination node, $loc_{curr}(d)$ the current location of n_d, $loc_{prev}(d)$ the previous location of n_d, $TTL\left(loc_{prev}(d)\right)$ the time-to-live of the previous location of n_d, and $type(d)$ the mobility type (mobile, stationary) of n_d. Ants are only used to collect and seminate location information to the routing tables of the nodes. The actual routing is then calculated on the location entries using a shortest-path algorithm.

T-ANT was developed in 2006 by Selvakennedy et al. [51]. It uses ants for cluster head election in *WSNs*. The algorithm uses elements from *Scalable Ant-based Routing* [38] described before (see Section 3.5). The actual routing is performed using a greedy routing scheme.

Ant-aggregation was developed by Misra and Mandal in 2006 [52] for use in *WSNs*. Here, ants are not used for routing but to determine the optimal in-network data-aggregation points (the optimal nodes for data-aggregation in the *WSN*). The algorithm is based on *Ant Colony Optimization*.

4. Conclusions

The versatile and dynamic nature of Wireless Multi-Hop Networks requires routing algorithms that are robust, adaptable, and scalable. Ant Algorithms are inspired from the self-organizing foraging behavior of natural ants, which show an incredible ability for robustness, adaptation and scalability despite being based on a set of simple mechanisms. In this chapter, we have first reviewed the seminal ant routing algorithm developed for routing in such networks, *AntHocNet*. In the following, we investigate more specific algorithms: *ARA* is a simple version of an ant colony optimization approach for routing in *MANETs*. *ARAMA* is targeted at *MANETs* and *WSNs* and focuses on fair energy use between the nodes of the network. *EEABR* is another algorithm focusing on energy efficiency, providing a more fine-grained selection of routing mechanisms. *DDCHA* is a data-centric protocol which divides a sensor field into subnets of nodes within communication range. *AMQR* is a routing algorithm for *MANETs* that extends upon *ARA*. A concept for dual-priority traffic, together with a notion of energy and latency constraints, is reported in the *PPRA* algorithm. To match requirements of different traffic types, we have also reviewed approaches using colored pheromones – here the colored pheromones form separate routing layers representing different route properties such as latency, jitter, or bandwidth. Two representatives of this class of algorithms are *MACO* and *SANETs* have been reviewed in this chapter.

Ant-inspired algorithms can be successfully applied for routing in Wireless Multi-Hop Networks, but due to the difficulty of the problem and different requirements priorities among Wireless Multi-Hop Networks, they have not converged into s single solution. Instead, we are facing an increasing number of algorithms and protocols following this idea. A short insight into this is given in the section on other ant-based algorithms for Wireless Multi-Hop Networks. Considering the constant change of sensor network hardware and software together with probably slightly different requirements, we are expecting this trend to continue and foresee new ant routing algorithms in the future.

Acknowledgements

The publication of this chapter was supported by the University of Klagenfurt.

Author details

Martina Umlauft[1] and Wilfried Elmenreich[2]*

1 Lakeside Labs, Klagenfurt, Austria

2 Institute of Networked and Embedded Systems, Alpen-Adria-Universität Klagenfurt, Klagenfurt, Austria

*Address all correspondence to: wilfried.elmenreich@aau.at

IntechOpen

References

[1] F. Zhao and L. Guibas. Wireless Sensor Networks: An Information Processing Approach. Morgan Kaufmann, 2014.

[2] Y. Toor, P. Muhlethaler, A. Laouiti, and A. D. La Fortelle. Vehicle ad hoc networks: Applications and related technical issues. IEEE Communications Surveys Tutorials, 10(3):74–88, 2008.

[3] V. Rastogi, V. J. Ribeiro, and A. D. Nayar. Measurements in olpc mesh networks. In 2009 7th international symposium on Modeling and optimization in Mobile, Ad Hoc, and Wireless Networks, pages 1–6, 2009.

[4] freifunk.net. http://start.freifunk.net/, last visited: Dec. 2010.

[5] FunkFeuer Wien – Verein zur Förderung freier Netze (ZVR: 814804682). 0xFF – FunkFeuer Free Net. http://funkfeuer.at/, last visited: Dec. 2010.

[6] W. Elmenreich and H. de Meer. Self-organizing networked systems for technical applications: A discussion on open issues. In J.P.G. Sterbenz. K.A. Hummel, editor, Proceedings of the Third International Workshop on Self-Organizing Systems, pages 1–9. Springer Verlag, 2008.

[7] Charles E. Perkins. Ad Hoc Networking. Addison Wesley, 2001. ISBN 0-201-30976-9.

[8] Ian F. Akyildiz, Xudong Wang, and Weilin Wang. Wireless mesh networks: A survey. Computer Networks, 47(4), March 2005.

[9] David B. Johnson. Routing in ad hoc networks of Mobile hosts. In proceedings of the workshop on Mobile computing systems and applications, pages 158–163, Santa Cruz, CA, December 1994. IEEE Computer Society.

[10] David B. Johnson, David A. Maltz, and Yih-Chun Hu. The Dynamic Source Routing Protocol for Mobile Ad hoc Networks (DSR). Ietf internet draft, IETF, July 2004. draft-ietf-manet-dsr-10.txt, work in progress.

[11] Charles E. Perkins and Pravin Bhagwat. Highly dynamic destination-sequenced distance-vector routing (dsdv) for mobile computers. In ACM SIGCOMM'94 Conference on Communications Architectures, Protocols and Applications, pages 234–244, 1994.

[12] Charles E. Perkins and Elizabeth M. Royer. Ad hoc on-demand distance vector routing. In Proceedings of the 2nd IEEE Workshop on Mobile Computing Systems and Applications, pages 90–100, New Orleans, LA, February 1999.

[13] Philippe Jacquet, Paul Muhlethaler, Thomas Clausen, Anis Laouiti, Amir Qayyum, and Laurent Viennot. Optimized Link State Routing Protocol for Ad hoc Networks. In INMIC 2001, Pakistan, 2001.

[14] Thomas Clausen and Philippe Jacquet, editors. Optimized Link State Routing Protocol (OLSR). IETF, October 2003. IETF Experimental RFC 3626.

[15] David Culler, Deborah Estrin, and Mani Srivastava. Overview of sensor networks. IEEE computer, pages 41–49, august 2004. Special Issue in Sensor Networks.

[16] Ian F. Akyildiz, Weilian Su, Yogesh Sankarasubramaniam, and Erdal Cayirci. A survey on sensor networks. IEEE Communications Magazine, pages 102–114, August 2002.

[17] Kay Rˮomer and Friedemann Mattern. The Design Space of Wireless

Sensor Networks. IEEE Wireless Communications, 11(6):54–61, December 2004.

[18] Salem Hadim and Nader Mohamed. Middleware Challenges and Approaches for Wireless Sensor Networks. IEEE Distributed Systems Online, 7(3), 2006. art. no. 0603-o3001.

[19] S. M. Hedetniemi, S. T. Hedetniemi, and A. L. Liestman. A survey of gossiping and broadcasting in communication networks. Networks, 18:319–349, 1988.

[20] W. R. Heinzelman, J. Kulik, and H. Balakrishnan. Adaptive Protocols for Information Dissemination in Wireless Sensor Networks. In ACM MobiCom'99, pages 174–185, Seattle, WA, USA, 1999.

[21] Brad Karp. Geographic Routing for Wireless Networks. PhD thesis, Harvard University, Cambridge, MA, USA, October 2000.

[22] Brad Karp and H. T. Kung. Greedy Perimeter Stateless Routing for Wireless Networks. In Sixth Annual ACM/IEEE International Conference on Mobile Computing and Networking (MobiCom 2000), Pages 243–254, Boston, MA, USA, August 2000.

[23] W. R. Heinzelman, A. Chandrakasan, and H. Balakrishnan. Energy-Efficient Communication Protocol for Wireless Microsensor Networks. In 33rd Annual Hawaii International Conference on System Sciences, pages 1–10, January 2000.

[24] K. Sohrabi, J. Gao, V. Ailawadhi, and G. J. Pottie. Protocols for self-Organization of a Wireless Sensor Network. IEEE Personal Communications, 7(5):16–27, October 2000.

[25] C. Intanagonwiwat, R. Govindan, and D. Estrin. Directed Diffusion: A Scalable and Robust Communication Paradigm for Sensor Networks. In ACM/IEEE International Conference on Mobile Computing and Networking (MobiCom'00), pages 56–67, Boston, MA, USA, 2000.

[26] Yan Zhang, Jijun Luo, and Honglin Hu, Editors. Wireless Mesh Networking, Architectures, Protocols and Standards. Auerbach Publications, 2007. ISBN 0-8493-7399-9.

[27] Jim Hauser, Dennis Baker, and W. Steven Conner. Draft PAR for IEEE P802.11 ESS Mesh. Technical report, IEEE, January 2004. IEEE P802.11 Wireless LANs, document 11-04/ 0052r2.

[28] S. Goss, S. Aron, J. L. Deneubourg, and J.M. Pasteels. Self-organized shortcuts in the argentine ant. Naturwissenschaften, 76:597–581, 1989.

[29] M. Dorigo and T. Stützle. Ant Colony Optimization. A Bradford Book, The MIT Press, 2004.

[30] M. Dorigo, V. Maniezzo, and A. Colorni. The ant system: Optimization by a Colony of cooperating agents. IEEE Transactions on Systems, Man, and Cybernetics-Part B, 26(1):1–13, 1996.

[31] M. Dorigo and L. M. Gambardella. Ant Colony system: A cooperative learning approach to the traveling salesman problem. IEEE Transactions on Evolutionary Computation, 1(1):53–66, April 1997.

[32] G. D. Caro and M. Dorigo. AntNet: Distributed Stigmergy control for communications networks. Journal of Artificial Intelligence Research (JAIR), 9:317–365, 1998.

[33] Gianni Di Caro, Frederick Ducatelle, and Luca Maria Gambardella. Anthocnet: an adaptive nature-inspired algorithm for routing in mobile ad hoc networks. European Transactions on

Telecommunications, 16(5):443–455, 2005.

[34] Mesut Günes,, Udo Sorges, and Imed Bouazizi. ARA – The Ant-Aolony Based Routing Algorithm for MANETs. In Stephan Olariu, editor, International Workshop on Ad Hoc Networking (IWAHN 2002), Pages 79–85, Vancouver, British Columbia, Canada, August 18–21 2002. IEEE Computer Society Press.

[35] Osama H. Hussein, Tarek N. Saadawi, and Myung J. lee. Probability routing algorithm for Mobile ad hoc networks. Journal in selected areas in communications, 23(12):2248–2259, 2005. IEEE.

[36] Y. Abdelmalek, O. Hussein, and T. Saadawi. A Simplified Receiver Assisted Routing Enhancement (RARE) in MANET. In BIONETICS, Budapest, Hungary, December 10–13 2007.

[37] Langgui Liu and Guangzeng Feng. A Novel Ant Colony Based QoS-Aware Routing Algorithm for MANETs. In L. Wang, K. Chen, and Y. S. Ong, editors, Advances in Natural Computation (ICNC 2005), Lecture notes in computer science, LNCS 3612, pages 457–466. Springer, 2005.

[38] Yoshitaka Ohtaki, Naoki Wakamiya, Masayuki Murata, and Makoto Imase. Scalable ant-based routing algorithm for ad-hoc networks. IEICE Transactions on Communications, 89(4):1231–1238, 2006. ISSN 0916-8516.

[39] D. Subramanian, P. Druschel, and J. Chen. Ants and reinforcement learning: A case study in routing in dynamic networks. Technical Report tr96-259, Rice University, July 1998.

[40] C. A. Santivánez, R. Ramanathan, and I. Stavrakakis. Making link-state routing scale for ad hoc networks. In 2nd ACM International Symposium on Mobile ad hoc networking and computing, pages 22–32, October 2001.

[41] Paul Barom Jeon and George Kesidis. Pheromone-Aided Robust Multipath and Multipriority Routing in Wireless MANETs. In 2nd ACM Intl. Workshop on Performance Evaluation of wireless ad hoc, sensor and ubiquitous networks (PE-WASUN), pages 106–113, Montreal, Quebec, Canada, October 10–13 2005.

[42] Tiago Camilo, Carlos Carreto, Jorge Sá Silva, and Fernando Boavida. An energy-efficient ant-based routing algorithm for wireless sensor networks. In Marco Dorigo, Luca Maria Gambardella, Mauro Birattari Alcherio Martinoli, Riccardo Poli, and Thomas Stützle, editors, 5th Intl. Workshop on ant Colony optimization and swarm intelligence (ANTS), lecture notes in computer science, LNCS 4150, pages 49–59, Brussels, Belgium, September 4–7 2006. Springer.

[43] Kwang Mong Sim and Weng Hong Sun. Multiple Ant-Colony Optimization for Network Routing. In Proc. 1st Intl. Symposium on Cyber Worlds (CW'02), pages 277–281, 2002.

[44] Thomas Halva Labella and Falko Dressler. A Bio-Inspired Architecture for Division of Labour in SANETs. In Proc. 1st IEEE ACM Intl. Conference on Bio-Inspired Models of Network, Information and Computing Systems (BIONETICS'06), Cavalese, Italy, Dec. 2006.

[45] Yuan-yuan Zeng and Yan-xiang He. Ant Routing Algorithm for Mobile Ad-hoc Networks Based on Adaptive Improvement. In IEEE Wireless Communications Networking and Mobile Computing (WiCOM), volume 2, pages 678–681, Wuhan, China, September 23–26 2005.

[46] Ming Liu, Yange Sun, Rui Liu, and Xiaoyan Huang. An Improved Ant

Colony QoS Routing Algorithm Applied to Mobile Ad Hoc Networks. In IEEE Wireless Communications, Networking and Mobile Computing (WiCOM), pages 1641–1644, Shanghai, China, September 21–25 2007.

[47] Shivanajay Marwaha, Chen Khong Tham, and Dipti Srinivasan. Mobile Agents based Routing Protocol for Mobile Ad Hoc Networks. In IEEE Global Telecommunications Conference (GLOBECOM'02), volume 1, pages 163–167, Taipei, Taiwan, Nov. 17-21 2002.

[48] Ying Zhang, Lukas D-Kuhn, And Markus P. J. Fromherz. Improvements on ant routing for sensor networks. In Marco Dorigo, Mauro Birattari, Christian Blum, Luca M. Gambardella, Francesco Mondada, and Thomas Stützle, editors, 4th Intl. Workshop on ant Colony optimization and swarm intelligence (ANTS), lecture notes in computer science, LNCS 3172, pages 154–165, Brussels, Belgium, September 5–8 2004. Springer.

[49] Daniel Cˆamara and Antonio Alfredo F. Loureiro. A Novel Routing Algorithm for Ad Hoc Networks. In 33rd Hawaii International Conference on System Sciences, Hawaii, January 4–7 2000.

[50] Daniel Câmara and Antonio Alfredo F. Loureiro. A novel routing algorithm for hoc networks. Baltzer Journal of Telecommunications Systems, 18(1–3):85–100, 2001. Kluwer Academic Publishers.

[51] S. Selvakennedy, S. Sinnappan, And Yi Shang. T-ANT: A nature-inspired data gathering protocol for wireless sensor networks. Journal of communications, 1(2):22–29, may 2006. Academy Publisher.

[52] Rajiv Misra and Chittaranjan Mandal. Ant-aggregation: Ant Colony Algorithm for optimal data aggregation in Wireless Sensor Networks. In IFIP Conference on Wireless and Optical Communications Networks, Bangalore, India, April 11-13 2006.

Chapter 5

An Innovative Maintenance Scheduling Framework for Preventive, Predictive Maintenance Using Ant Colony Optimization

Abhishek Kaul

Abstract

The fourth industrial revelation has brought exponential technologies in the area of digitization, internet of things (IOT), artificial intelligence (AI), and optimization which has helped mining companies to increase the availability and utilization of equipment's. As mining companies implement predictive maintenance technologies, to improve overall equipment availability, there is more value to be unearthed if predictive maintenance is optimized with the production schedules. Ant colony optimization (ACO) is a metaheuristic that is inspired from the behavior of real ants to solve combinatorial optimization problems. This chapter describes an innovative maintenance scheduling framework in the context of optimizing schedules for preventive maintenance and predictive maintenance, with multiple constraints for optimized dynamic schedule to reduce the maintenance time, and production losses.

Keywords: ant colony optimization, predictive maintenance, preventive maintenance, mining, schedule, optimizing mining equipment schedule

1. Introduction

The high and volatile commodity prices are caused by unanticipated changes demand and supply [1]. These volatile prices put cost pressure on mining organization to optimize operations. The availability and utilization of mining equipment's, is the major contributor for an organization to manage costs and supply disruptions.

Traditionally, maintenance activity for mining equipment, relies on a series of time based or equipment running hours based checks for scheduling maintenance activities. The fourth industrial revolution provide organizations with a balanced approach to reduce costs with safety. As new technologies get deployed, the operations and maintenance landscape is continually being digitized by mechanization, automation, industrial internet of things (IIoT) and IT-OT (information technology – operational technology) integration. These technologies provide visibility to real time operations data. Analysing the data with artificial intelligence (AI) adds the ability to predict and respond to operational disruptions, for example – predict the next failure date of the asset and provide perspective guidance for maintenance. Further, optimization adds the capability to synchronizing the scheduled, predicted maintenance activity with production schedule to minimize the maintenance costs and production losses.

Ant colony optimization (ACO) [2, 3] is a metaheuristic for solving hard combinatorial optimization problems. This was proposed by Dorigo et al., inspired from the behaviour of real ants, which use pheromones as a communication medium to find the shortest path to food from the colony. Analogous to the biological example, ACO is based on indirect communication within a colony of simple agents, called (artificial) ants, mediated by (artificial) pheromone trails. In ACO algorithms there are several generations of artificial ants which search for good solutions. In each generation, each ants find a solution by going step by step through many probabilistic decision till a solution is found. Ants that found good solutions put some amount of pheromone on the edges of path to mark their path. This will help attract the next generation of ants to find solutions near the good space. Generally pheromone values of ants are guided by the specific heuristic that is used for evaluating decisions.

In this chapter, we will focus on the framework for optimizing the preventive maintenance, predictive maintenance and production schedule. The first section will cover the maintenance strategies and framework. The second section will give a brief overview of the ACO. The third section will cover the maintenance solution framework for mining equipment's. In the last section we will conclude our recommendations for mine equipment maintenance scheduling.

2. Maintenance strategies and framework

The systematic, optimally sequenced activities and framework, through which mining companies can sustainably manage their equipment, its performance, risks and operating expenses over their entire life cycle, for the purpose of achieving its organizational objectives and plan, can be defined as an enterprise asset management. Mining equipment maintenance requires series of checks by the equipment operator, for better diligence apart from unscheduled fixes. The frequency is dependent on the combination of equipment performance and the running hours in a specified time interval.

Maintenance costs can be upto 30% of direct costs [4] and more in terms of operations disruptions. To control maintenance costs, mining organization have centred their efforts on areas such as optimizing scheduled maintenance operations, deferring nonessential maintenance, reducing maintenance manpower, controlling inventories of spare parts even more adequately, and using contract maintenance support [5]. Below all the maintenance types, which are performed by the maintenance team at varied frequency depending on the nature of inputs from various sources are described

2.1 Regular maintenance—Type I

The first daily checks and safety practices are mandated by the OEM and occur the start of every shift or during operator changes. These proactive checks involve the inspection of all major system parameters such as engine temperature, tyre pressures, oil levels and control surfaces and are performed at the mine site itself. These checks can be part of the total productive maintenance (TPM) strategy and incorporates activities and actions performed by equipment operators with the intention of to ensure failure-free operations, fewer breakdowns and efficiency [6].

2.2 Preventive maintenance—Type II

Routine planned maintenance is about avoiding, reducing or eliminating the consequences of failures. The frequency of performance-based maintenance is done primarily on performance (running hours, KM run) of the equipment, time based (weekly, monthly, quarterly, yearly, and multiple of years) or as recommended by OEM and updated time to time.

The determination of inspection intervals is based on the reliability level to prevent potential catastrophic failures that augments extensive maintenance and increased downtime costs. The insights are considered to develop a preventive maintenance plan that is based on the deterioration of equipment key components and renewed life after repair [7]. The mechanical parts deterioration factors depend on multiple factors which are wear and tear, corrosion, operational fatigue and weather, and operator skills. This deterioration is a continuous process which is time and usage dependent [8].

2.3 Breakdown or corrective maintenance—Type III

The unscheduled breakdown has a major impact in both – production and expedited maintenance costs. Furthermore, the ability to fix it right first time is fundamental to ensure equipment is back to operation quickly. The core objective for repairing the equipment is in less time, with higher accuracy and support maintenance technicians to diagnose the failure properly is key for improved utilization. The breakdown and corrective maintenance require timely availability of spares and management of spares just in time is one of the main challenges of delay in maintaining of the equipment.

Condition based maintenance (CBM) provides the insights of the equipment component, either real time or at a specific frequency to analyse the condition for effective decision making for dynamic corrective maintenance schedule for restoration. CBM insights can be utilized either for preventive or corrective maintenance depending on the organizational maintenance strategy [9].

2.4 Overhaul and shutdown maintenance—Type IV

This type of maintenance is mostly capital in nature and performed at relatively large intervals. It may be done for several reasons due to severe break- down, accident or to overhaul the entire equipment to enhance the useful life of the equipment. This overhaul or shutdown maintenance is managed as a project with procurement of spare parts and maintenance activities are scheduled and synchronized as per the timeline agreed upon. Overhaul is complete check and review of the mining equipment and depends on the performance of the equipment or between the mid to end of useful life to enhance the useful life of the equipment. This type of maintenance is performed mostly on the large equipment for example, dragline and shovels. This maintenance can run up-to multiple months and are managed as a project with a sequence of activities and schedules, to minimize equipment downtime.

2.5 Predictive maintenance—Type V

This type of maintenance is performed based on accurate prediction when an equipment or any of its components are going to fail. If the prediction attribute is derived the maintenance can be executed just before such failure is predicted to occur.

There are systems like vehicle health monitoring system (VHMS) which provide, frequency-based sample data about the equipment performance for example, running hours, speed, rpm, load, engine temperature, payload so on and so forth. Similarly, weather related data is collected through weather application programming interface (API), which helps to understand the impact of ambient temperature, precipitation, humidity etc. on the equipment performance. The main idea behind the IIoT is to connect computers, devices, sensors, and industrial equipment and applications within an organization and to continually collect data, such as system errors and machine telemetry, from all of these with the aim of analysing and acting on this data in order to optimize operational efficiencies. Predictive maintenance is more effective than performing preventive maintenance at frequent intervals, which could also be costlier because unnecessary maintenance may be applied on equipment.

The above types of maintenance strategies help organization to develop a comprehensive maintenance framework to maximize value and realise benefits. Specifically for mining equipment, the maintenance planning can be done at

- Workshop—typically for moving equipment like dump trucks, haulers, dozers, graders, and wheel loaders

- Onsite at mine—typically for large slowly/non-movable, equipment like dragline, large shovels

2.6 Strategies and framework

Overall, maintenance affects, all aspects of business efficacy, safety, environmental impact, energy efficiency, product quality, customer service, plant availability and cost. Many times scheduled maintenance activity is seldom integrated with the production [10] which leads to unplanned production losses due to planning of scheduled and preventive maintenance activities.

Therefore, the selection of the right maintenance framework plays a significant role in preserving the functions of the equipment and supporting mining organization value drivers:

- Improve revenue—increased asset availability and greater reliability in line with production schedule i.e. grow more revenue from the same asset base

- Reduce operating costs and expenses—more timely and precise interventions, increase asset life, less downtime, high utilization

In order to achieve these benefits, efficient combination of preventive (Type II) and predictive maintenance (Type V) with production schedule is required to reduce the overall maintenance execution time and maximize production.

3. Understanding ant colony optimization

Ant colony optimization (ACO) was inspired by the observation of the behaviour of real ants. Real ants, which use pheromones as a communication medium to find the shortest path to food from the colony [11]. As in the case of real ants, the problem is to find the food, in the case of artificial ants, it is to find a good solution to a given optimization problem.

One ant (either a real or an artificial one) can find a solution to its problem, but only cooperation among many individual ants through stigmergy enables them to find good solutions [12]. In real ants stigmergic communication happens via the pheromone that ants deposit on the ground. Artificial ants live in a virtual world, hence they only modify numeric values (called for analogy artificial pheromones) associated with problem states they visit while building solutions to the optimization problem. Real ants simply walk, choosing a direction based on local pheromone concentrations and a stochastic decision policy. Artificial ants also create solutions step by step, moving through available problem states and making stochastic decisions at each step.

The ACO metaheuristics has an initialization step and then a loop over three basic components. In one iteration of the loop, there are steps to construct the solution by all ants, improve (optional) the solutions with local search also and then an update of the pheromones.

Algorithm for ant colony optimization metaheuristic

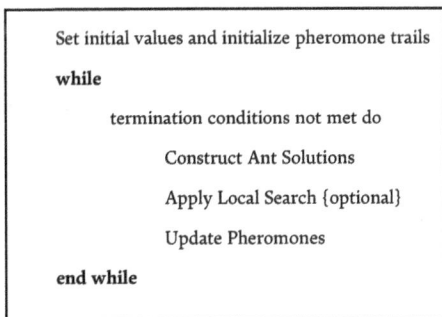

```
Set initial values and initialize pheromone trails
while
        termination conditions not met do
            Construct Ant Solutions
            Apply Local Search {optional}
            Update Pheromones
    end while
```

In the next section ACO is explained using travelling salesman problem example.

3.1 Example: the traveling salesman problem

Travelling salesman problem (TSP) can be easily applied to the Ant colony optimization. In this problem, there are a set of locations (cities) where the travelling salesman has to visit. The key constraints are to visit each location and visit only once. The distance between cities (locations) are given and the objective is to find the shortest distance between them.

In the example below, there are four cities, c1, c2, c3 and c4. The lengths of the edges between vertices is proportional to the distance between cities i.e. c13 is the distance between city 1 and city 3.

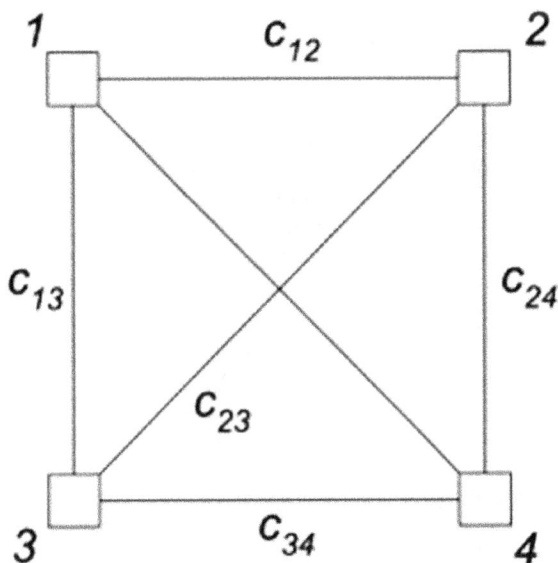

The pheromone is associated with the edges of the graph. Each ant starts from a randomly selected city and then at each construction step it moves along the edges of the graph. An ant chooses probabilistically the edge to follow among the available ones (those that lead to yet unvisited vertices).

3.1.1 Sample equation for implementation

$$p(cij|s^p) = \frac{\lambda^\alpha_{ij} \cdot \eta(c_{ij})^\beta}{\sum_{c_{ij} \in N(s^p)} \lambda^\alpha_{ij} \cdot \eta(c_{ij})^\beta c}, \forall c_{ij} \in N(s^p). \qquad (1)$$

where pheromone value associated with the component cij is λij. Function that assigns at each construction step a heuristic value to each feasible solution component $c_{ij} \in N(s^p)$ is $\eta(\cdot)$ which is commonly called heuristic information. Positive parameters, whose values determine the relative importance of pheromone versus heuristic information are α and β.

The solution is constructed once the ant has visited all the vertices of the graph. When all the ants have constructed the solutions by visiting the vertices of the graph, pheromone levels on the edges are updated positively for good solutions and reduced for the bad solutions. The update function, typically does two things one is to increase the pheromone values for set of good solutions and second is to reduce the pheromone value by implementing an evaporation function. This helps to avoid rapid convergence of the algorithm and helps in the exploration of new areas.

3.1.2 Sample equation for pheromone update

$$\lambda_{ij} \leftarrow (1-\rho)\tau_{ij} + \rho \sum_{s \in S_{upd}|c_{ij} \in s} F(s) \qquad (2)$$

where set of solution that are used for update are S_{upd}, the parameter that is called for evaporation rate is $\rho \in (0,1]$, and $F : S \rightarrow R+0$ a function such that $f(s) < f(s') => F(s) \geq F(s'), \forall s \neq s' \in S$. $F(\cdot)$ is commonly called the fitness function.

Ant colony optimization has been shown to perform quite well on the TSP [13].

3.2 Other applications of ACO: Scheduling problems

ACO has been used for many applications including scheduling problem, vehicle routing problem, assignment problem, set problem, device sizing problem in nanoelectronics physical design, antennas optimization and synthesis, image processing. In this chapter our focus is on the scheduling problems.

In scheduling problems, jobs have to be processed on one or many machines such that some objective function is optimized. For these problems the following is true (a) the processing time of jobs is known beforehand and (b) processes of jobs cannot be interrupted. Typically the construction graph for scheduling problems is represented by the set of jobs (for single-machine problems). Some of the key research papers published for scheduling problems are:

- Group-shop scheduling problem (GSP) [14]

- Sequential ordering problem (SOP) [15]

- Job-shop scheduling problem (JSP) [16]

- Multistage flowshop scheduling problem (MFSP) with sequence dependent setup/changeover times [17]

- Permutation flow shop problem (PFSP) [18]

- Open-shop scheduling problem (OSP) [19, 20]

- Single-machine total tardiness problem with sequence dependent setup times (SMTTPDST) [21]

- Single machine total tardiness problem (SMTTP) [22]

- Resource-constrained project scheduling problem (RCPSP) [23]

- Single machine total weighted tardiness problem (SMTWTP) [24–26]

Out of the above scheduling applications, the SMTWTP has the best application for our maintenance schedule. In the subsequent sections, SMTWTP is described in greater detail for developing the optimal schedule for maintenance of mining equipment's.

4. Maintenance solution framework

The optimized maintenance scheduling framework is recommended to be built using ant colony optimization model using structured content found in a typical maintenance ecosystem.

The first step is to predict the failure based on survival analysis, with certain confidence level before its occurrence. The next step is to combine the predictive maintenance and preventative maintenance schedule using optimization model. This optimization model determines which equipment should be assigned to which day in the maintenance workshop bay for minimizing waiting time, maximizing production and ultimately increase the availability.

Maintenance Solution Framework

Fleet of trucks   Framework   Workshop

In the solution framework, two types of maintenance activities are considered: first preventive maintenance which is based on the recommendation of OEM. The second maintenance is predictive maintenance which is based on probabilistic failure. The next section will cover the derivation of predictive maintenance schedule.

4.1 Predictive maintenance schedule

The predictive maintenance schedule is derived from Cox regression model which gives survival probability distribution function. The Cox regression model is shown in below

$$H(t) = H_0(t) \exp\left(\beta 1 Y1 + \beta 2 Y2 + \cdots + \beta n Yn\right) \tag{3}$$

Where the expected hazard is $H(t)$ at time t, the baseline hazard is $H_0(t)$ and it represents the hazard when all of the independent variables) $Y1, Y2, \ldots Yn$ when they are equal to zero. Based on the collected data the model estimates $\beta 1, \beta 2, \ldots \beta n$.

The expected hazard function increases as the days progress. This function is converted to survival days and remaining useful life (RUL). RUL is defined as the duration left for the occurrence of breakdown based on the probability threshold of failure i.e. how many days when the cumulative probability falls below 60%. The RUL has been extensively used in calculating the reliability-based research in the mine system to derive the occurrence of the failure, so the appropriate action can be taken proactively

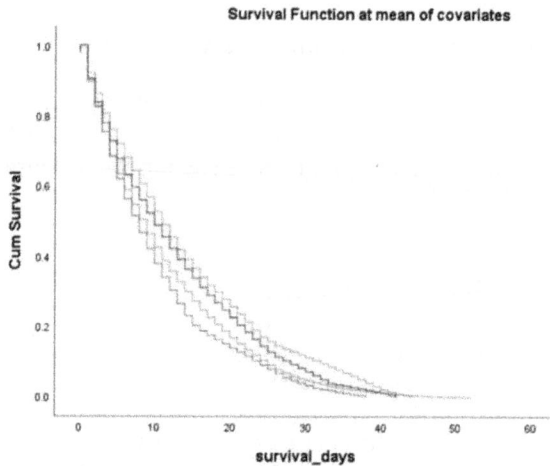

Survival Function at mean of covariates

This survival data (predictive failure day) is used in combination with the preventive maintenance, production schedule and other constraints to optimizing the maintenance schedule.

4.2 Optimization of maintenance schedule

The optimization of maintenance schedule requires to determine the optimal maintenance day for each truck in a time horizon so that the maintenance time, production loss is minimised while meeting the preventative maintenance schedule requirements and minimizing the probability of failure (predictive maintenance – survival days). In the above sections we discussed on the application of ACO in the context of SMTWTP.

In order to formulate the problem using SMTWTP, it is assumed that there is one workshop and it has one bay for carrying out maintenance activities (single machine). One truck is represented as one job and a fleet has n trucks which need to be scheduled for maintenance. Based on historical analysis, the time for each job i.e. time for carrying out maintenance activities is available (processing time p_j). The due date (d_j) of processing is provided by the preventive maintenance schedule for each job (truck). The completion time of job j is defined as C_j. The earliness of job is defined as E_j, if the job is completed early and the tardiness of job is defined as T_j, if the job is completed late. The probability of failure (f_j) at C_j is provided by the predictive maintenance schedule (RUL). The cost functions (w) for earliness and lateness take the probability of failure (f_j) into consideration.

The objective is to find the truck fleet maintenance scheduling sequence that minimizes the function as below.

$$1|d_j| \sum_{-j} w\, E_j + \bar{w}_j T_j \qquad (4)$$

where

d_j – due date of job (preventative maintenance scheduled day for truck)

$\underset{-j}{w}$ – unit cost of earliness i.e. cost of maintenance too early based

$E_j = \max\{0, d_j - C_j\}$; earliness of job

$T_j = \max\{0, C_j - d_j\}$; tardiness of job

$\bar{w}_j$ – unit cost of tardiness i.e. cost of lost production if failure before maintenance (f_j)

$N = \{1,...,n\}$; n trucks (jobs) have to be sequentially processed (1 job = 1 truck maintenance activities) at workshop (1 bay)

This function can be modelled to minimizes the total weighted earliness tardiness (z) as below

$$z = \min \sum_{j \in N} \underset{-j}{w}\, E_j + \bar{w}_j T_j \qquad (5)$$

Mainly there are three key requirements of the ACO algorithm.

- A construction graph – The construction graph consist of C components for the n jobs that need to be assigned at the optimal positions. In the graph each points is connected by L arcs.

- Problem constraints – The main constraint is that all the jobs have to be scheduled and scheduled only once.

- Update pheromone trails – This refer to the attractiveness of scheduling or assigning the job j to position i.

Applying the ACO algorithm, in the initialization step, a colony A of m ants is generated, where each ant corresponds to a random feasible solution. The next is the iterative step where the acquired knowledge (pheromone level) is fetched and job assignment attractiveness is calculated. Next the ant generations are merged and only the best ants are retained for the optimal solution. Lastly pheromone evaporation and deposit are updated and the process continues till the maximum number of Generations are reached.

```
ACO pseudo code
Input parameters

  • N, is a set of n trucks (jobs that need to be processed in workshop)

  • C, the number of colonies

  • n, the number of ants in the colony (i.e. size)

Output solution
```

- A (near) global optimum S* of cost z_a.

Steps

1. Setup

   1. Initialize the generation counter g=0

   2. Create an initial colony A of size n

   3. Set the best solution S* to the ant a $\in$ A with the least weighted earliness tardiness

   4. Initialize the pheromone level $\rho(0)$ using a subset of the colony A, and set $\rho(1) = \rho(0)$

2. Loop Step

   1. Set g=g+1

   2. Build colony of ants while taking into account for knowledge acquiredρ (g) and attractivenessη (g), and apply a dynamic visibility function

   3. Merge colonies of generations g and (g−1), and retain the best n ants

   4. Update optimal solution S*

   5. Update the pheromone level $\rho(g+1)$ by accounting for the evaporation and deposit

3. Stopping Criterion

   1. If g<C, then go-to Step 2.

At the end of step 2.3 to further enhance quality of the solution i.e. the retained n ants and speed up the convergence towards near optimal solution, a local search criteria can be applied. Hybrid approaches with local search criterial include beam search [20], scatter search, tabu search [27], threshold accepting [28], and neighbourhood search [29]. These search criteria help to efficiently guide the ants movements towards global optima. In the paper by M'Hallah and Alhajraf [30] ant colony systems for the single-machine total weighted earliness tardiness scheduling problem, they provide empirical evidence of using variable neighbourhood search (VNS) to improve the overall quality of the retained ants and converge towards a near global optimum.

By applying ACO to SMTWTP, the total cost for early or late maintenance is reduced by optimally assigning the truck to the workshop for maintenance activities based on the preventive maintenance schedule and predicted maintenance (RUL).

5. Conclusion

In this chapter, the importance of optimally planning maintenance activities for mining organization was discussed. A solution framework for optimizing the preventive maintenance, predictive maintenance and production schedule was proposed using ant colony optimization. Many mining organization can benefit by using this solution framework to reduce the overall maintenance costs and production losses.

As mining companies adopt and implement Industry 4.0, this maintenance solution framework has the potential to evolve beyond maintenance schedules to allocation of ore to customer demand, planning truck routing, and even to mine planning. ACO as part of the wider Swarm intelligence algorithms presents the capacity to achieve Industry 4.0 vision, where individual machines cooperate through self-organization, that is, without any form of central control to achieve the organization KPIs.

Conflict of interest

The views expressed in this chapter are my own and are not representative of my employer.

Author details

Abhishek Kaul
IBM Consulting, Singapore

*Address all correspondence to: abhishekkaul@gmail.com

IntechOpen

References

[1] Commodity Price Volatility. Available from: https://treasury.gov.au/sites/default/files/2019-03/01_Commodity_price_volatility.pdf [Accessed: 27 December 2021]

[2] Dorigo M, Di Caro G. The ant colony optimization meta-heuristic. In: Corne D, Dorigo M, Glover F, editors. New Ideas in Optimization. London, UK: McGraw Hill; 1999. pp. 11-32

[3] Dorigo M, Di Caro G, Gambardella LM. Ant algorithms for discrete optimization. Artificial Life. 1999;5(2):137-172

[4] Henderson K, Pahlenkemper G, Kraska O. Integrated asset management—An investment in sustainability. Procedia Engineering. 2014;83:448-454. DOI: 10.1016/j.proeng.2014.09.077

[5] Unger EJ. An Examination of the Relationship Between Usage and Operating and Support Costs for Air Force Aircraft. Santa Monica, CA: RAND Corp.; 2007

[6] Brodny J, Tutak M. Application of elements of TPM strategy for operation analysis of mining machine. In: IOP Conference Series: Earth and Environmental Science journal. Vol. 95. IOP Publishing; 2017. p. 42019. Available from: https://iopscience.iop.org/article/10.1088/1755-1315/95/4/042019

[7] Angeles E, Kumral M. Optimal inspection and preventive maintenance scheduling of mining equipment. Journal of Failure Analysis and Prevention. 2020:1-9. DOI: 10.1007/s11668-020-00949-z. Available from: https://www.researchgate.net/publication/343413414_Optimal_Inspection_and_Preventive_Maintenance_Scheduling_of_Mining_Equipment.

[8] Changyou L, Haiyang L, Song G, Yimin Z, Zhenyuan L. Gradual reliability sensitivity analysis of mechanical part considering preventive maintenance. Advances in Mechanical Engineering. 2014;6:829850

[9] Marseguerra M, Zio E, Podofillini L. Condition-based maintenance optimization by means of genetic algorithms and Monte Carlo simulation. Reliability Engineering & System Safety. 2002;77(2):151-165

[10] Liao W, Pan E, Xi L. Preventive maintenance scheduling for repairable system with deterioration. Journal of Intelligent Manufacturing. 2010;21(6): 875-884

[11] Deneubourg J-L, Aron S, Goss S, Pasteels J-M. The self- organizing exploratory pattern of the argentine ant. Journal of Insect Behavior. 1990;3:159

[12] Dorigo M, Socha K. An introduction to ant colony optimization. IRIDIA—Technical Report Series. 2006

[13] Stützle T, Dorigo M. ACO algorithms for the traveling salesman problem. In: Miettinen K, Mäkelä MM, Neittaanmäki P, Périaux J, editors. Evolutionary Algorithms in Engineering and Computer Science. New York: Wiley; 1999. p. 163

[14] Blum C. ACO applied to group shop scheduling: A case study on intensification and diversification [permanent dead link]. In: Proceedings of ANTS 2002. Vol. 2463 of Lecture Notes in Computer Science. 2002. pp. 14-27

[15] Gambardella LM, Dorigo M. An ant colony system hybridized with a new local search for the sequential ordering problem. INFORMS Journal on Computing. 2000;12(3):237-255

[16] Martens D, De Backer M, Haesen R, Vanthienen J, Snoeck M, Baesens B. Classification with Ant Colony Optimization. IEEE Transactions on Evolutionary Computation. 2007;**11**(5): 651-665

[17] Donati AV, Darley V, Ramachandran B. An ant-bidding algorithm for multistage flowshop scheduling problem: Optimization and phase transitions. In: Advances in Metaheuristics for Hard Optimization. Springer; 2008. pp. 111-138. Available from: https://www.springerprofessional.de/en/an-ant-bidding-algorithm-for-multistage-flowshop-scheduling-prob/2812100

[18] Stützle T. An ant approach to the flow shop problem. Technical Report AIDA-97-07. 1997

[19] Pfahring B. Multi-agent search for open scheduling: Adapting the ant-Q formalism. Technical Report TR-96-09. 1996

[20] Blem C. Beam-ACO, hybridizing ant colony optimization with beam search. An application to open shop scheduling. Technical Report TR/IRIDIA/2003-17. 2003

[21] Gagné C, Price WL, Gravel M. Comparing an ACO algorithm with other heuristics for the single machine scheduling problem with sequence-dependent setup times. Journal of the Operational Research Society. 2002;**53**: 895-906

[22] Bauer A, Bullnheimer B, Hartl RF, Strauss C. Minimizing total tardiness on a single machine using ant colony optimization. Central European Journal for Operations Research and Economics. 2000;**8**(2):125-141

[23] Merkle D, Middendorf M, Schmeck H. Ant colony optimization for resource-constrained project scheduling. In: Proceedings of the Genetic and Evolutionary Computation Conference (GECCO 2000). 2000. pp. 893-900. DOI: 10.5555/2933718. 2933886

[24] den Besten M. Ants for the single machine total weighted tardiness problem [master's thesis]. University of Amsterdam; 2000. Available from: https://www.academia.edu/9009128/An_Ant_Colony_Optimization_Application_to_the_Single_Machine_Total_Weighted_Tardiness_Problem

[25] den Bseten M, Stützle T, Dorigo M. Ant colony optimization for the total weighted tardiness problem. In: Proceedings of PPSN-VI, Sixth International Conference on Parallel Problem Solving from Nature. Vol. 1917 of Lecture Notes in Computer Science. 2000. pp. 611-620. DOI: 10.5555/645825.669098

[26] Merkle D, Middendorf M. An ant algorithm with a new pheromone evaluation rule for total tardiness problems. In: Real World Applications of Evolutionary Computing. Vol. 1803 of Lecture Notes in Computer Science. 2000. pp. 287-296

[27] Huang KL, Liao CJ. Ant colony optimization combined with taboo search for the job shop scheduling problem. Computers & Operations Research. 2008;**35**:1030-1046

[28] Marimuthu S, Ponnambalam SG, Jawahar N. Threshold accepting and ant-colony optimization algorithms for scheduling m-machine flow shops with lot streaming. Journal of Materials Processing Technology. 2009;**209**: 1026-1041

[29] Behnamian J, Fatemi Ghomi SMT, Zandieh M. Development of a hybrid meta heuristic to minimise earliness and tardiness in a hybrid flow shop with sequence-dependent setup times. International Journal of Production Research. 2010;**48**(5):1415-1438

[30] M'Hallah R, Alhajraf A. Ant colony
systems for the single-machine total
weighted earliness tardiness scheduling
problem. Journal of Scheduling. 2015;**19**.
DOI: 10.1007/s10951-015-0429-x.
Available from: https://www.
researchgate.net/publication/
275220942_Ant_colony_systems_for_
the_single-machine_total_weighted_
earliness_tardiness_scheduling_problem

www.ingramcontent.com/pod-product-compliance
Lightning Source LLC
Chambersburg PA
CBHW081239190326
41458CB00016B/5839